L Rohnd

Conservation and
the Gospel of Efficiency

Conservation and the Gospel of Efficiency

The Progressive Conservation Movement, 1890–1920

SAMUEL P. HAYS

University of Pittsburgh Press

Published 1999 by the University of Pittsburgh Press, Pittsburgh, Pa. 15261
Originally published 1959 by Harvard University Press

Printed in the United States of America
10 9 8 7 6 5 4 3 2

ISBN 0-8229-5702-7

To
Frederick Merk

Preface to the Paperback Edition

Over a decade has now passed since this book was first published; a reprinting in paperback form provides an opportunity to view it in the perspective of those years. I am happy to take advantage of this opportunity for two reasons. First, many readers assumed erroneously that the primary significance of the book was in its treatment of the evolution of conservation policies in the Progressive Era. On the contrary, I had hoped that the work would turn the reader away from the substance of conservation as such and into the realm of political structure. I was concerned not so much with the idea of efficiency, which many have focused on, but with the political structure and system of decision-making which efficiency and all that it involved represented. My concern was not intellectual history, but the history of the structure of power in modern America. For this reason my subsequent research and writing on voting behavior, reform in municipal government in the Progressive Era, and community-society patterns in the same period are simply extensions of the major theme of the current book rather than diversions. They explore precisely the same subject matter as does this work on conservation. But while I left this implication to speak for itself in the former edition, here I wish to take advantage of the revision to make it explicit.

I would be the first to admit that the implications as elaborated in this preface are far more highly developed than they would have been in 1957. As one plays with the idea of modernization, its ramifications and implications become far more elaborate, extensive, and explicit. Yet the fact remains that at this vantage point in time my subsequent remarks can bring out more effectively the implications of the original work which were not at the time fully seen. My later explicit concern for social research and political structure is implicit in this work; an intellectual autobiography would have to acknowledge as much.

Rather than traverse this ground in detail, I wish here simply to sketch the larger implications of ideas which were most succinctly stated originally in the first and last chapters. The natural tendency of the reader is to fit the book into his pigeon-hole of "natural resource policy." Perhaps by this explicit statement it can be transferred to the pigeon-hole of "political structure" and I can thereby draw attention more sharply to the type of problem which I consider to be the most crucial in the evolution of modern, urban-industrial America.[1]

Conservation cannot be considered simply as a public policy, but, far more significantly, as an integral part of the evolution of the political structure of the modern United States. This twist in approach to the study of resource policy, though difficult to grasp, is fundamental. To most historians the significance of the conservation movement lies in the substance of conservation programs, sustained-yield forestry, multiple-purpose river development, and efficient public land management; in the active political forces supporting and opposing those policies; and in the relative success of their implementation in Congress and the administrative agencies. But this book is concerned in a broader way not with events and decisions, but with political structure; not with the way in which political forces produced a given outcome, but in which events and outcomes can shed light on the forces which constitute the larger system. We shall work not from forces to results but from events back into the patterns of forces.

This book, therefore, constitutes an attempt to re-create the larger political structure of the Progressive Era. For this purpose the decisions and events of conservation are simply an occasion for the examination of political change, a prism through which more extensive developments are observed. For too long historians have been mesmerized with events, with the episodic and the isolated, and have failed to be concerned with pattern. Here we shift from event to structure so that the more significant context can come into view and

[1] For more extended statements of this problem see Samuel P. Hays, "The Politics of Reform in Municipal Government in the Progressive Era," *Pacific Northwest Quarterly* (October 1964), 55 #4, 157-169, and Hays, "Political Parties and the Community-Society Continuum," in William N. Chambers and Walter Dean Burnham (eds), *The American Party Systems* (Oxford, 1967).

reveal history as a web of human interrelationships, of patterns of human interaction within the larger society rather than a simple sequence of beads on a string. To explore the larger context of this structure is the goal of examining in detail the events of resource policy. But first let us examine more precisely the framework of political structure within which conservation in the Progressive Era had meaning.

The dynamics of conservation, with its tension between the centralizing tendencies of system and expertise on the one hand and decentralization and localism on the other, is typical of a whole series of similar tensions between centralization and decentralization within modern American society. The poles of the continuum along which these forces were arrayed can be described briefly. On the one hand many facets of human life were bound up with relatively small-scale activities focusing on the daily routines of job, home, religion, school and recreation in which a pattern of inter-personal relationships developed within relatively small geographical areas. Relying on personal interaction and often personal communication, often giving rise to a sense of consciousness and experience limited by place and locality, these contexts readily generated self-conceptions of protecting and developing a limited segment of social, economic and political life as against the larger world "out there." On the other hand, however, modern forms of social organization gave rise to larger patterns of human interaction, to ties of occupation and profession over wide areas, to corporate systems which extended into a far-flung network, to impersonal media of communication and impersonal—statistical—forms of understanding, to the reliance upon expertise and to centralized manipulation and control. Although Americans as a whole were swept on toward "modern" life, they remained at different stages in the process and the tension of decentralization against centralization remained persistently strong.

The direction of change was toward centralization. To many people the external characteristics of this process—efficiency, expertise, order—constituted the spirit of "progressivism." The process took two major forms, systematization and functional organization. Systematization emphasized the integration of human activity into more tightly knit patterns of regular human interaction. Some of these, such as corporate systems, both private and public, were consciously directed ventures, planned to bring together many resources into one activity for pre-determined ends. Some grew out of the daily choices of millions of men and women to participate in the conditions of life brought about by greater system: patronage at large-scale supermarkets rather than the corner drug store; participation in the mass media of metropolitan newspapers and radio rather than the local press with its emphasis on daily life in the small community; the drive for education and the acquisition of skills through which incomes could be derived for higher standards of living. Efficiency, expertise, system infused the entire order.

Functional organizations, which grew rapidly, played an equally important role in ordering human relationships. Their roots lay in specialization, the concentration on more specialized aspects of the work process as a basis of occupational differentiation, and the equally significant concentration upon the acquisition of specialized skills to enable one to play a more adequate role in his special task. Specialized functions gave rise to the establishment of ties with others on the basis of common functions and common outlooks. The bricklayer, faced with an employment situation specific to bricklayers, joined with fellow bricklayers for common action to influence the conditions which affected them; their organizations extended out to encompass the entire nation. And, likewise, it was for the doctor and lawyer, the teacher and minister, the grocer and jeweler, the beef cattle producer and the tomato grower. A host of contacts arose which grew into organizations variously called trade unions, trade associations and commodity organizations. Because the network of economic life in which they found themselves extended in ever wider circles, so did their organizations and the conditions which they wished to influence. And to exercise that influence they were drawn toward the larger centers of decision-making and power—the state and federal governments.

These new forms of organization tended to shift the location of decision-making away from the grass roots, the smaller contexts of life, to the larger networks of human interaction. This upward shift can be seen in many specific types of development: the growth of city-wide systems of executive action and representation in both school and general government to supersede the previous focus on ward representation and action; the similar upward shift in the management of schools and roads from the township to the county and the state department of public instruction and the state highway commission; the upward shift in regulation of economic life from the state to the federal regulatory agency. These upward shifts did not arise out of new political theories or the inherent logic of a proper distribution of governmental powers, but rather from the fact that those who fashioned the new patterns of system and functional organizations sought a framework of decision-making consistent in scope and applicability with the scope of affairs they wished to control. It reflected an upward shift in the level of human life and interaction which permeated the entire society. New contexts of human life had arisen, giving rise to new contexts of conditions to be controlled. Control now became a more elaborate process, involving measurement and prediction, reliance upon the experts who could develop and manipulate information, and techniques for shaping the course of events to reach predictable outcomes.

But only portions of the political order were actively involved in this process of modernization. Many geographical areas, many people, many communities lay outside the center of the new tendencies. In fact, they found themselves being drawn into them, often reluctantly, at times enticed by material advantages which modernization brought to them personally, but fearful of the consequences and especially the persistent loss of control over the conditions which affected them. They reached out for the benefits of modernization, but often fought the implications for decision-making it embodied. Often they accepted without protest because of the benefits which came. In the city, community political life at the ward level declined and social organization at the community level shifted from civic involvement in innovation in public affairs to passive acceptance of benefits and the establishment of family-related institutions. Rural areas struck back with more force; they bitterly resisted, for example, the establishment of state highway commissions which shifted control from traditional local officials and they equally resisted the consolidation of schools which transferred the context of educational and other aspects of social life to larger, more mysterious, less understandable and less controllable centers. By the same token, although the South and the West sought assistance from those large-scale sources of capital, technical knowledge and entrepreneurship—both private and public—which lay at the center of national corporate life in the North and East, they bitterly resented the control from outside which accompanied it.

The thrust of modernization and the resulting tension between centralizing and decentralizing forces took place in different geographical contexts. This can be observed within the confines of cities and urban government, or within the context of broader regions and state governments. But it also took place within the larger context of private and public national life. The growth of private corporations and resistance to them, the centralization of the process of distribution accompanied by opposition from independent merchants, the evolution of the role of government in railroad regulation to a single system of direction in the United States Railroad Administration with its attendant hostility from shippers, the interaction between federal and state programs to regulate private business such as pure food and drug laws: in each case the larger forces of economic life, with a scope far broader than cities, regions or states, sought a national, uniform context of action and a central point of decision-making which greatly limited the political variables to be controlled.

Natural resource policies played an integral role in these processes at the national level. From the point of view of the larger context of historical, social and political change, they involved both the extension of the new techniques of modernization—system, expertise, centralized direction and manipulation—and the activation of tensions between centralizing and decentralizing forces. Conservationists were led by people who promoted the "rational" use of resources, with a focus on efficiency, planning for future use, and the application of expertise to broad national problems. But they also promoted a system of decision-making consistent with that spirit, a process by which the expert would decide in terms of the most efficient dove-tailing of all competing resource users according to criteria which were considered to be objective, rational, and above the give-and-take of political conflict. In short, they sought to substitute one system of decision-making, that inherent in the spirit of modern science and technology, for another, that inherent in the give-and-take among lesser groupings of influence freely competing within the larger system.

Examination of the evolution of conservation political struggles, therefore, brings into sharp focus two competing political systems in modern America. On the one hand the spirit of science and technology, of rational system and organization, shifted the location of decision-making continually upward so as to narrow the range of influences impinging upon it and to guide that decision-making with large, cosmopolitan considerations, technical expertness, and the objectives of those involved in the wider networks of modern society. These forces tended toward a more closed system of decision-making. On the other, however, were a host of political impulses, often separate and conflicting, diffuse and struggling against each other within the larger political order. Their political activities sustained a more open political system, in which the range of alternatives remained wide and always available for adoption, in which complex and esoteric facts possessed by only a few were not permitted to dominate the process of decision-making, and the satisfaction of grass-roots impulses remained a constantly viable element of the political order.

The conservation movement in the Progressive Era, therefore, sheds light not so much on the content of public policy but on the entire political structure, the types of human interaction, perspective and goals peculiar to the different portions of that structure, and the rival systems of decision-making which have developed in modern society. These elements of political structure constitute the most far-reaching aspect of modern American society; into them a review of natural resource politics in the Progressive Era can provide extensive insight.

Samuel P. Hays

Pittsburgh, Pennsylvania
June 8, 1968

Acknowledgments

I am deeply indebted to a great number of people for assistance while preparing this monograph. I was fortunate, indeed, to be among those many doctoral candidates who have worked under the direction of Professor Frederick Merk, and who have experienced his unfailing personal concern for students, his quiet wisdom, and his contagious enthusiasm for scholarly research. While Professor Merk contributed much in the way of inspiration, the Theodore Roosevelt Memorial Association helped, through its financial assistance in the form of the Theodore Roosevelt Memorial Fellowship, to carry the research for this book through its most vital period, the academic years 1950–51 and 1951–52.

In the task of finding and using manuscripts upon which this study is based I have received assistance from a great number of sources. The late Nora Cordingly, curator of the Theodore Roosevelt collection at Harvard University, and her successor, Thomas Little, went out of their way to make the facilities of that collection available. For equal readiness to assist I am indebted to the staff of the Manuscripts Division of the Library of Congress, including David C. Mearns, Robert H. Land, and C. Percy Powell; to the staff of the National Archives, and in particular Oliver W. Holmes, Chief Archivist of the Natural Resources Records Branch; and to Miss Emma Stephenson and Mrs. Zara Jones Powers of the Sterling Memorial Library at Yale University. Mr. Arthur B. Darling, curator of the Francis G. Newlands papers, kindly granted permission to explore this valuable collection; the Society for the Protection of New Hampshire Forests opened its early manuscripts for use; and the National Reclamation Association made available a memoir of George H. Maxwell. Over a dozen living heirs of those whose papers I have used have kindly granted permission to quote from the manuscripts. For all of these I am grateful.

The criticisms and counsel of a number of friends and colleagues have contributed in a less tangible, yet equally vital, manner. Among them are John M. Blum, Allan G. Bogue, Richard Current, Frank Freidel, William H. Matchett, and Herbert Rowen. Finally, an unmeasurable amount of assistance has come from a co-worker, Barbara Darrow Hays, who, through her knowledge of ecology, contributed much to the scientific aspects of the study, who joined with her typewriter in a summer of note-taking at the Library of Congress and the National Archives, and who has provided continued encouragement and faith that years of effort would eventually yield results.

S. P. H.

November 25, 1958

Contents

Abbreviations

When the following manuscript collections are cited in the footnotes, designations to precise boxes, files, or scrapbooks refer to labels on the material in the collection. Precise citations have not been made in the case of the Roosevelt manuscripts which are adequtely indexed, both chronologically and according to author and recipient of each letter.

RG #48	National Archives, Record Group #48, Records of the Department of the Interior, Correspondence of the Office of the Secretary
RG #49	National Archives, Record Group #49, Records of the General Land Office
RG #77	National Archives, Record Group #77, Records of the War Department, Army Section, Correspondence of the Office of the Chief of Engineers
RG #95	National Archives, Record Group #95, Records of the United States Forest Service
RG #107	National Archives, Record Group #107, Records of the War Department, Army Section, Correspondence of the Secretary of War
MLD	Myra Lloyd Dock MSS, Library of Congress
JRG	James Rudolph Garfield MSS, Library of Congress
GHM	George H. Maxwell MSS, National Archives, Record Group #115, Records of the Bureau of Reclamation
WJM	W J McGee MSS, Library of Congress
FN	Frederick H. Newell MSS, Library of Congress
FJN	Francis G. Newlands MSS, Sterling Memorial Library, Yale University, New Haven, Connecticut
GP	Gifford Pinchott MSS, Library of Congress
TR	Theodore Roosevelt MSS, Library of Congress and Widener Memorial Libary, Harvard University, Cambridge, Massachusetts
WHT	William Howard Taft MSS, Library of Congress

Conservation and the Gospel of Efficiency

Chapter I

Introduction

The conservation movement has contributed more than its share to the political drama of the twentieth century. A succession of colorful episodes—from the Pinchot-Ballinger controversy, through Teapot Dome, to the Dixon-Yates affair—have embellished the literature of the movement, and called forth fond memories for its later leaders. Cast in the framework of a moral struggle between the virtuous "people," and the evil "interests," these events have provided issues tailor-made to arouse the public to a fighting pitch, and they continue to inspire the historian to recount a tale of noble and stirring enterprise. This crusading quality of the conservation movement has given it an enviable reputation as a defender of spiritual values and national character. He who would battle for conservation fights in a worthy and patriotic cause, and foolhardy, indeed, is he who would sully his reputation by opposition!

Such is the ideological tenor of the present-day conservation movement and of its history as well. But, however much an asset in promoting conservation, this dramatic fervor has constituted a major liability in its careful analysis. For the moral language of conservation battles differed markedly from the course of conservation events. Examining the record, one is forced to distinguish sharply between rhetoric and reality, between the literal meaning of the terminology of the popular struggle and the specific issues of conservation policy at stake. Conservation neither arose from a broad popular outcry, nor centered its fire primarily upon the private corporation. Moreover, corporations often supported conservation policies, while the "people" just as frequently opposed them. In fact, it becomes clear that one must discard completely the struggle against corporations as the setting in which to understand conservation history, and permit an entirely new frame of reference to arise from the evidence itself.

Conservation, above all, was a scientific movement, and its role in history arises from the implications of science and technology in modern society. Conservation leaders sprang from such fields as hydrology, forestry, agrostology, geology, and anthropology. Vigorously active in professional circles in the national capital, these leaders brought the ideals and practices of their crafts into federal resource policy. Loyalty to these professional ideals, not close association with the grass-roots public, set the tone of the Theodore Roosevelt conservation movement. Its essence was rational planning to promote efficient development and use of all natural resources. The idea of efficiency drew these federal scientists from one resource task to another, from specific programs to comprehensive concepts. It molded the policies which they proposed, their administrative techniques, and their relations with Congress and the public. It is from the vantage point of applied science, rather than of democratic protest, that one must understand the historic role of the conservation movement.[1]

The new realms of science and technology, appearing to open up unlimited opportunities for human achievement, filled conservation leaders with intense optimism. They emphasized expansion, not retrenchment; possibilities, not limitations. True, they expressed some fear that diminishing resources would create critical shortages in the future. But they were not Malthusian prophets of despair and gloom. The popular view that in a fit of pessimism they withdrew vast areas of the public lands from present use for future development does not stand examination. In fact, they bitterly opposed those who sought to withdraw resources from commercial development. They displayed that deep sense of hope which per-

[1] A recent statement of a contrary point of view is J. Leonard Bates, "Fulfilling American Democracy: The Conservation Movement, 1907 to 1921," in *Mississippi Valley Historical Review* (June 1957), 44, 29–57.

vaded all those at the turn of the century for whom science and technology were revealing visions of an abundant future.

The political implications of conservation, it is particularly important to observe, grew out of the political implications of applied science rather than from conflict over the distribution of wealth. Who should decide the course of resource development? Who should determine the goals and methods of federal resource programs? The correct answer to these questions lay at the heart of the conservation idea. Since resource matters were basically technical in nature, conservationists argued, technicians, rather than legislators, should deal with them. Foresters should determine the desirable annual timber cut; hydraulic engineers should establish the feasible extent of multiple-purpose river development and the specific location of reservoirs; agronomists should decide which forage areas could remain open for grazing without undue damage to water supplies. Conflicts between competing resource users, especially, should not be dealt with through the normal processes of politics. Pressure group action, logrolling in Congress, or partisan debate could not guarantee rational and scientific decisions. Amid such jockeying for advantage with the resulting compromise, concern for efficiency would disappear. Conservationists envisaged, even though they did not realize their aims, a political system guided by the ideal of efficiency and dominated by the technicians who could best determine how to achieve it.

This phase of conservation requires special examination because of its long neglect by historians. Instead of probing the political implications of the technological spirit, they have repeated the political mythology of the "people versus the interests" as the setting for the struggle over resource policy. This myopia has stemmed in part from the disinterestedness of the historian and the social scientist. Often accepting implicitly the political assumptions of eliteism, rarely having an axe of personal interest to grind, and invariably sympathetic with the movement, conservation historians have considered their view to be in the public interest. Yet, analysis from outside such a limited perspective reveals the difficulty of equating the particular views of a few scientific leaders with an objective "public interest." Those views did not receive wide acceptance; they did not arise out of widely held assumptions and values. They came from a limited group of people, with a particular set of goals, who played a special role in society. Their definition of the "public interest" might well, and did, clash with other competing definitions. The historian, therefore, cannot understand conservation leaders simply as defenders of the "people." Instead, he must examine the experiences and goals peculiar to them; he must describe their role within a specific sociological context.

This study, then, is an examination of the ideas and values of conservation leaders as a special group in American society, and an analysis of the wider implications of their attempt to work out the concept of efficiency in resource management. I will not undertake here an exhaustive treatment of the specific struggles over conservation policy. I am more concerned with the development of a group of ideas determining the behavior of technicians, and of the impact of these ideas upon the wider public. I will make a special effort to illuminate the points of cooperation and conflict between the values of technology implicit in conservation and competing values with which they came into contact. This study, it is hoped, will bring the history of the conservation movement into sharper focus, and emphasize its hitherto neglected features. It might also serve as a fruitful point of departure for exploring the wider political implications of technology in recent American history.

Chapter II

Store the Floods

The modern American conservation movement grew out of the firsthand experience of federal administrators and political leaders with problems of Western economic growth, and more precisely with Western water development. Such men as Frederick H. Newell of the United States Geological Survey, George H. Maxwell, a California water law specialist, Representative Francis G. Newlands of Nevada, and President Theodore Roosevelt joined to promote a federal irrigation program.[1] Their experience in this campaign, and their later experiences—as they constructed and operated irrigation works—with problems of water rights, speculation, and siltation gave rise to extensive ideas about water conservation. These views gradually became crystallized into an over-all approach and by 1908 emerged as a concept of multiple-purpose river development. The movement to construct reservoirs to conserve spring flood waters for use later in the dry season gave rise both to the term "conservation" and to the concept of planned and efficient progress which lay at the heart of the conservation idea.[2]

[1] There is no comprehensive history of irrigation in the United States. The best general accounts are George Wharton James, *Reclaiming the Arid West* (New York, 1917); Elwood Mead, *Irrigation Institutions* (New York, 1903); Frederick Haynes Newell, *Irrigation in the United States* (New York, 1906); William E. Smythe, *The Conquest of Arid America* (New York, 1905); and Ray P. Teele, *Irrigation in the United States* (New York, 1915).

[2] The term conservation is used in this sense, for example, in United States Department of the Interior, *10th Annual Report of the United States Geological Survey, 1889–90*, part II, *Irrigation*, 22–23; in *Bradstreet's*, August 31, 1902 (FN #16, Clippings, v. 3); and in *The Forester* (Nov. 1898), 4, 223.

New Horizons in Water Use

In their task of gathering technical data about stream flow, hydrographers of the United States Geological Survey evolved the idea that water is a single resource of many potential uses.[3] This simple reorientation in outlook opened up new vistas of water development. It became the fundamental idea in water conservation.

In 1888 Congress authorized the first water resources investigation of the arid lands, a measure which Major John Wesley Powell, Chief of the Survey, had encouraged for over a decade.[4] Under this law the hydrographic branch of the Survey set out to measure water supplies, locate reservoirs and canals, and map areas susceptible of irrigation. It soon turned to studies of the movement of ground water and sedimentation, and before long was called upon to expand its work to the East.[5] On the basis of this information federal officials planned Western irrigation works. Corporations interested in irrigation, water power, and domestic water supply also drew upon the new data.[6] These private groups, in fact, encouraged the ever-

---

Others used the term in connection with "the preservation and conservation of the range" and the "conservation" of water supplies in forested areas. See, for example, *Proceedings, American National Livestock Association, 1899*, 176, and *Forestry and Irrigation* (Apr. 1907), 13, 204. These uses, however, were not widespread until 1907.

[3] A brief summary of the history of the Geological Survey is in Institute for Government Research, *Service Monographs of the United States Government No. 1, The U. S. Geological Survey* (New York, 1918). The annual reports of the Survey, beginning in 1890, provide a more detailed account.

[4] Powell's views are in 45th Congress, 2nd Session, *House Executive Document 73*, 25–45.

[5] This work can be followed in detail in the successive annual reports of the U. S. Geological Survey, beginning in 1890. A good account of water resources investigation in the Appalachian Mountains is in Frederick H. Newell, "Forests and Water-Power," address before the Cotton Manufacturers' Association, printed in the *Manufacturers' Record*, May 12, 1914 (FN #17, Clippings, v. 4).

[6] For examples, see *The Forester* (July 1904), 10, 292–294, and the *Troy Times*, May 11, 1907 (FN #16, Clippings, v. 6).

widening activities of the Survey's hydrographic branch and supported its campaign for larger congressional appropriations.[7]

A young engineer, Frederick Haynes Newell, took charge of this work. Born in Bradford, Pennsylvania, in 1862, Newell graduated from the Massachusetts Institute of Technology in 1885 as a mining engineer. Three years later he became the assistant hydraulic engineer of the United States Geological Survey, and the first man assigned to carry out the Act of 1888. In 1890 he was promoted to chief hydrographer. From the very start of his official career, Newell took an active interest in the scientific work of the federal government and especially in promoting the dissemination of scientific information. He served, for example, as voluntary secretary of the National Geographic Society in 1892–93 and again in 1897–99. He also promoted a federal water development program, at first for irrigation, but later for power, navigation, and flood control as well. During the Roosevelt administration Newell became one of the architects of water policy and of the entire conservation movement.

The Geological Survey, among federal agencies, did not pioneer in water resource studies; yet it did bring forth a wider concept of water use. In 1824 Congress instructed the Army Corps of Engineers to improve the navigable streams, and since that date the Corps had carried out frequent hydrographic investigations such as the extensive Humphreys and Abbott survey of the Mississippi River completed in 1866.[8] Confined by Congress to the improvement of navigation, however, the Corps limited its hydrographic work to measurements of low water flow. It placed upon private landowners the responsibility for collecting data about drainage and floods, even though such matters, in the same watershed, intimately affected navigation. Until the advent of the Geological Survey these wider uses of water remained uninvestigated. As the California Commission on Public Works remarked in 1895, "The Army Engineers . . .

[7] See, for example, George C. Warner to Brigadier-General Alexander Mackenzie, March 31, 1908 (RG #77, #63743).

[8] A convenient summary of the Corps' water activities is in W. Stull Holt, *The Office of the Chief of Engineers of the Army; its non-military history, activities and organization* (Baltimore, 1923).

failed to appreciate the importance of the study of the water resources of the country. . . . It was left to the United States Geological Survey, through its Hydrographic . . . Branch, to collect the much-needed information."[9]

The Corps of Engineers also held a narrow view of water use and water development. Viewing rivers primarily in terms of transportation, the Corps confined its congressional reports to the effect of new projects on navigation.[10] It referred to water power, irrigation, and drainage as secondary to navigation; it did not propose studies or plans for the development of all possible uses of water. The Corps of Engineers, commented Mr. Carl Grunsky, a leader in the civil engineering profession, "has never looked upon the related problems in the broad progressive way that led the U. S. Geological Survey into a study of stream flow."[11] At the same time, the Corps regarded the Geological Survey as a competitive administrative agency and sought to protect its own role in water development by resisting coordination of navigation with any other water use.

The scientists and engineers of the Geological Survey approached water development from a fresh point of view, unhampered by limited interests or institutional loyalties. They investigated flood water as part of a cycle of precipitation, evaporation, percolation, run-off, and stream flow, rather than as simple quantities to be diverted or as instruments of navigation. They were as concerned with the sediment content and mineral quality of water as with its

[9] *Report of Commission of Public Works to the Governor of California, 1895*, 138, as quoted by Carl E. Grunsky in C. E. Grunsky, H. M. Chittenden, and H. F. Labelle, "The Flood of March, 1907, in the Sacramento and San Joaquin River Basins, California," *Proceedings, American Society of Civil Engineers* (Apr. 1908), 34, 368.

[10] *Ibid.* "Without data relating to the river in all its stages," Grunsky also commented, "the U. S. Engineer Corps will be but poorly equipped to combat or acquiesce in the recommendations of engineers studying drainage problems."

[11] Carl E. Grunsky, "Presidential Address before the San Francisco Association of members of the American Society of Civil Engineers, February 16, 1912," MSS in FGN, Waterways—River Regulation, Correspondence, 1912, 1.

physical movement.[12] This approach gave rise to a broader view of river planning. In federal programs, the Survey argued, all possible uses of water should be considered so that rivers could produce the greatest possible benefit for man. Multiple-purpose river-basin development in later years arose directly from the experiences and ideas of these new hydrographers in the Geological Survey.

The Federal Government Undertakes Irrigation

While Newell and his field force carried on their hydrographic studies, Western leaders undertook a search for capital for reservoir construction which was to bring the federal government directly into the task of water development. After the irrigation boom-and-bust of 1887–93, private investors turned away from the West to seek more lucrative opportunities. The West, in turn, began to look to the federal government for aid. The Carey Act of 1894, passed in response to this demand, sought to solve the problem by granting a million acres of land to each Western state to be used to finance irrigation. This program produced few projects, so that by the late 1890's the West, through the National Irrigation Congress, demanded a new program.[13] The Act of 1888, which initiated hydrographic studies, had anticipated direct federal financing, and members of the Geological Survey, especially Frederick H. Newell, strongly backed the proposal.[14] Toward the end of the nineties these federal officials joined with Western irrigators to promote a program of federal investment in irrigation.

George H. Maxwell, a young California lawyer, spearheaded this campaign. A native of Sonoma, California, Maxwell became a court stenographer, developed an interest in irrigation, studied law, was admitted to the bar in 1882, and became a specialist in California

[12] An excellent expression of this point of view is in WJ McGee, "Water as a Resource," in "Conservation of Natural Resources," *Annals of the American Academy of Political and Social Science* (May 1909), 33, 37–50.

[13] By 1910 seven states had developed 1,089,677 acres under the terms of the Carey Act; 742,618 of this total were in Idaho. See Teele, *Irrigation*, 67.

[14] Newell wrote *Irrigation in the United States* (New York, 1902), primarily to further the cause.

water law. In the late 1890's he became convinced that irrigation could solve national social problems by decentralizing population from urban centers back to the land. In 1907, for example, he formed the Homecroft Society to popularize homesteads for urban workers on the fringes of industrial centers. He also became the major irrigation propagandist in the country. He led the educational campaign for federal financing, and, having achieved this goal, vigorously entered the fight for a comprehensive federal program for multiple-purpose river development. A tireless and devoted worker who secured little financial reward for his efforts, Maxwell was impelled by the sheer conviction that homes on the land would save the country from a great peril.[15]

In 1896 Maxwell led the advocates of federal financing to victory in the annual convention of the National Irrigation Congress, when they persuaded that organization to back their proposal.[16] It was more difficult for him to arouse the enthusiasm of the rest of the nation for the plan. Early in 1898 he set out to gain the backing of the nation's commercial and industrial interests by convincing them that more irrigation would increase Western farm population and enlarge markets for Eastern business.[17] Responding to these pleas with enthusiasm, the National Board of Trade, the National Business Men's League and the National Association of Manufacturers passed resolutions at their annual conventions in 1898 in support of federal aid to irrigation, and continued to do so each year until the National Reclamation bill became law.[18] In 1899 Maxwell organized the National Irrigation Association, located at Chicago; from here he disseminated literature to newspapers and to the general public and published a monthly periodical, *Maxwell's Talisman*.[19] Together with the annual resolutions of the National

[15] An unpublished memoir of Maxwell is in the files of the National Reclamation Association in Washington, D.C.

[16] William E. Smythe, *The Conquest of Arid America* (2nd ed., New York, 1905), 272–273.

[17] *Ibid.*, 273.

[18] *The Forester* (Nov. 1900), 6, 273, for example.

[19] *Maxwell's Talisman*, published from 1902 through 1908, and again in 1912 and 1913 contains a running account of Maxwell's activities and point of view.

Irrigation Congress, this campaign aroused sufficient public sentiment to persuade both major parties in 1900 to adopt platform planks which called for federal construction of irrigation works.

While Maxwell conducted his educational campaign, several Westerners promoted their cause in Congress. Representative Francis G. Newlands of Nevada offered the specific proposal which Congress accepted. Newlands had long taken an interest in irrigation. Although not a native-born Westerner—he was born in Mississippi in 1848—he went to California to practice law in 1870, soon after graduating from Yale and Columbia. When his father-in-law, William Sharon, wealthy silvermine owner and Senator from Nevada, died in 1889, Newlands moved to Nevada to manage the estate; from that time on he became deeply involved in the state's economic and political affairs. In 1892 he was elected to Congress, where he served in the House until 1903, and then in the Senate until his death in 1919. Here he played a leading role in the fight for federal irrigation and for over a decade labored unsuccessfully to persuade Congress to adopt a multiple-purpose river development program for the entire nation.[20]

Convinced that irrigation farming would provide the only remedy for Nevada's declining population, Newlands plunged into the task of promoting that cause soon after he moved to the state.[21] At his own expense he investigated possible reservoir sites on Nevada rivers, and presented his findings to the public in a pamphlet published in 1891.[22] Fearing that speculators might acquire these sites, he purchased several and offered to sell them to any water users' association for their original price plus interest charges.[23] Newlands played a leading role in the first National Irrigation Congress in 1891. During the mid-1890's he concentrated on the silver question, as a leading advocate of bimetalism; but as the West once more

[20] The best account of the public life of Francis G. Newlands is A. B. Darling (ed.), *The Public Papers of Francis G. Newlands* (2 vols., 1932).

[21] *Proceedings, National Irrigation Congress, 1900*, 114–115.

[22] Francis G. Newlands, "An Address to the People of Nevada on Water Storage and Irrigation" (Reno, 1891) (FGN, Irrigation, 1891–1900).

[23] Newlands to the Secretary of the Interior, June 16, 1904 (FGN, Scrapbook #13).

turned to irrigation problems toward the end of the decade, so did he.

In 1901 Newlands proposed that the federal government finance irrigation through a Reclamation Fund composed of proceeds from the sale of Western public lands.[24] Thus, the West would pay for its own development. The Secretary of the Interior, he advocated, should have complete discretion in selecting projects for construction and in apportioning funds to each. The Nevada Representative hoped to forestall any possibility that disagreements among Western congressmen over the location of projects might retard the entire program. Congressional control of annual appropriations, he argued, would produce the same inefficiency, confusion, and delay prevalent in rivers and harbors work. Newlands hoped that expert knowledge and planning, rather than logrolling, would determine the course of federal irrigation construction.[25] This provision for considerable executive discretion in resource development and management became a central feature of the later multiple-purpose program and of the entire conservation movement.

Representative Newlands added two other important items to his proposals. Family, rather than corporation farmers, he argued, should benefit from federal irrigation. No individual should receive water rights for more than eighty acres of land from a federal project. On the other hand, he contended, the Secretary of the Interior should prevent speculation by having the authority to withdraw from all forms of entry land which might be included in the program.[26] These proposals arose, not from Newlands' democratic political convictions, but from his view that a growing farm population provided the best hope for the economic progress of Nevada

[24] See "Bartine MSS" in the Newlands MSS, Irrigation, History of National, 2. In the later scramble among politicians to receive credit for the Newlands Act, there was much confusion as to who first had proposed the Reclamation Fund. Most concerned denied any knowledge of its origin. For example, see *Albuquerque Daily Citizen*, July 19, 1904 (FGN, Scrapbook #16). The earliest mention of the Fund that I have found, however, is in the bill introduced by Newlands, Jan. 16, 1901, a copy of which is in Smythe, *Conquest*, 342–344.

[25] 60th Congress, 1st Session, *Congressional Record*, 395.

[26] Smythe, *Conquest*, 343.

and the entire Mountain West. A farm of eighty acres sufficed for irrigation agriculture. The more such homesteads the government could carve out of its projects, the greater the benefit for the West.[27] Moreover, Newlands intended that the anti-speculation proposal would protect the program not only from unscrupulous land companies but also from the speculative predispositions of the settlers themselves which he personally had faced in his Nevada transactions.

These views did not please many who, although desiring the aid of the federal government, feared its restrictions. Most Western congressmen, for example, demanded a larger limited acreage. Newlands complied by raising the maximum to 160 acres, the figure which Congress finally approved. Westerners also opposed the plan to withdraw irrigable lands from all forms of entry. Once before, under the Act of 1888, similar withdrawals had aroused sufficient antagonism to force their restoration to entry. The very word "withdrawal" aroused Western farmers to a fighting pitch. Heretofore they had faced withdrawals for railroad construction, withdrawals for forest reservations, and withdrawals for irrigation, and they had fought, often successfully, to restore these lands to the public domain. The West now hesitated to grant the Secretary of the Interior power to suspend entry even temporarily, for fear that the suspensions might become permanent.[28] Actual development of withdrawn lands might not occur until the far-distant future. Newlands compromised on this point, too; the final Act provided that areas in proposed irrigation projects be withdrawn from all private entry except under the homestead laws.

The Newlands proposal met stiff opposition from Eastern Republicans as well. Federal aid to irrigation, they argued, would create unfair Western competition with Eastern farmers.[29] The East mellowed considerably when Senator Thomas Carter of Montana

[27] *Proceedings, National Irrigation Congress, 1900*, 114–115.

[28] Newlands discussed these problems in a long letter to William F. Herrin, Feb. 5, 1905 (FGN, Letters).

[29] A summary of petitions from the East to Congress both for and against the measure is in *Forestry and Irrigation* (Feb. 1902), 8, 50–51, 70–71, 77; (Mar. 1902), 8, 134–136. A brief account of the arguments on both sides is in E. Louise Peffer, *The Closing of the Public Domain* (Stanford, 1951), 36–38.

counterattacked in the spring of 1901 by filibustering to death the Rivers and Harbors bill with its many projects close to the hearts of Eastern congressmen. The prodding of a new chief executive, enthusiastic about Western irrigation, also helped persuade the Republicans to yield. President McKinley had refused to push the measure in the face of Republican opposition, but Theodore Roosevelt eagerly championed the cause. Roosevelt could not increase Republican votes for the Newlands measure, but he did persuade the party leaders to permit the House to consider it, and it passed on June 17, 1902.[30] Following the Newlands plan, it established the Reclamation Fund and gave the Secretary of the Interior authority to select and construct projects.

Roosevelt contributed even more to the irrigation movement by publicly identifying himself with it. Having lived in the semi-arid West, he had first-hand knowledge of its vital need for water. During his Western campaign tour in 1900 he reiterated the Republican platform pledge and emphasized his great personal interest in a federal irrigation program. Shortly after the campaign, in a letter to the National Irrigation Congress he repeated his support of the cause.[31] Both Gifford Pinchot, Chief of the Forestry Bureau, and Frederick H. Newell influenced Roosevelt's views on the subject even before the election of 1900. These two men were overjoyed at the opportunity which Roosevelt's advance to the presidency in 1901 offered them to carry out their plans. The new president invited them to make suggestions for his first message to Congress. Following their advice Roosevelt made clear to the lawmakers that he personally supported the irrigation measure and disagreed strongly with the hostile Republican leaders.[32]

[30] Roosevelt, without success, appealed to Speaker Cannon to withhold opposition; see Roosevelt to Cannon, June 13, 1902 (TR). See also William E. Smythe, "Democracy and the West, the Newlands Irrigation Act," (FGN, Irrigation, History of National, 2), 20–24.

[31] Roosevelt to the 9th Irrigation Congress, Nov. 16, 1900, *Proceedings, National Irrigation Congress, 1900*, 104–108.

[32] Theodore Roosevelt, *Theodore Roosevelt, An Autobiography* (New York, 1913), 394-396; Gifford Pinchot, *Breaking New Ground* (New York, 1947), 189–190.

The new president's attitude toward the Newlands measure was not an isolated affair. He took a keen interest in all conservation matters and identified the entire movement as "my policy." This interest in conservation stemmed, in part, from Roosevelt's personal love for the out-of-doors, and, in part, from his admiration for organization and efficiency in economic affairs. In early life he became an avid big-game hunter, and helped to found an organization of like-minded people—the Boone and Crockett Club. As governor of New York in 1899 and 1900 he took a special interest in the Adirondacks forest reserve, in fish and game affairs, and in streamlining the state's resource administration. As president, Roosevelt originated few of the new conservation ideas, but he did give full rein to those officials in his administration who promoted efficient resource development, and freely lent his personal prestige to their cause.

Conservation Requires an Effective Water Law

President Roosevelt entrusted the administration of the Newlands Act to the Reclamation Service, a new branch of the Geological Survey, and placed Frederick H. Newell in charge. In 1907 the Service became an independent Bureau directly under the Secretary of the Interior. In 1903 the Secretary approved four projects and the same year the Service began its work by tackling the Roosevelt reservoir on the Salt River in Arizona. In 1905 it completed its first project, the Truckee-Carson ditch in Nevada; and by 1910 some twenty-four others were under way.[33] In carrying out this work those at the forefront of the federal irrigation movement faced a variety of new problems. Their experience with questions such as water rights, speculation, and silt control played a significant role in shaping the larger water development concepts of the Roosevelt Era.

From the day of its inauguration, for example, the Reclamation Service faced a confusing Western water law which greatly hampered its work. In the East, where rainfall was abundant, water

[33] An account of the progress of federal reclamation work is in James, *Reclaiming the Arid West.*

rights created few problems; riparian land owners could legally use water flowing through or by their land. The West, however, an area of slight precipitation where the demand for water far exceeded its supply, required different legal arrangements. Most Western states adopted the doctrine of prior appropriation of water rights. Anyone could establish a prior appropriation by posting notice at the point of diversion and filing with the county clerk a statement of the amount of water claimed. If he continued to use this water beneficially, a prior appropriator retained his title over all later claimants.[34]

This method of determining water rights often led to much confusion. If a prior appropriator found others using his supply, he sued to defend his rights. The court then reviewed the record to determine who had a prior claim. But no one knew how much water was available. Moreover, since the law provided that filings be recorded by counties instead of by watersheds, no one court could determine the total number of water rights claimed on a single stream. Consequently, judges often established rights far in excess of available supply. Invariably appealed to higher courts, litigation became extremely expensive. Moreover, a title established in one case secured the appropriator against only that one claimant. If others later challenged his title, he had to repeat the same costly litigation. Many, in fact, chose to use extralegal means to protect their rights. Such confusion retarded both private and public irrigation development. Private corporations would not risk funds on projects which involved insecure titles, and, before the federal government undertook construction in any state, it also demanded that water use on its projects be protected by more adequate state laws.[35] The Reclamation Service, in fact, detailed one of its officials, Mr. Morris Bien, to deal exclusively with this question. Bien drew up a model water law which the Reclamation Service, with some success, tried to persuade states to adopt.[36]

[34] Mead, *Irrigation Institutions*, 62–87. [35] *Ibid.*, 62–87.

[36] An excellent account of this problem is William E. Smythe, "The Battle in the States," *Out West* (August 1902), 17, 233–37. Bien described his own work in "Proposed State Code of Water Laws," *Proceedings, National Irriga-*

Western states gradually evolved a more orderly system. In Wyoming, for example, an appointed state engineer, after having determined the amount of water available, decided priorities, enforced them, and granted new rights. His administrative organization conformed to watersheds and his decisions were subject to court review. Few states, however, copied the Wyoming law, which the engineering profession considered to be a model statute.[37] The state engineer usually received far less power and in some cases reform consisted of only a complete judicial determination of existing priorities. The leader of Wyoming water law reform, Elwood Mead, became the state's first official engineer and for many years thereafter worked actively to introduce the Wyoming law into other Western states.[38] When in 1898 he became Chief of the Office of Irrigation Investigations in the Department of Agriculture he pursued this task on an even wider scale.[39]

Western water rights were originally established by local custom and later protected by state laws and local courts. Since Congress confirmed these arrangements in 1866, water rights, even on the public lands, were subject to state, rather than federal law. The federal irrigation program brought to the fore the potential conflicts between state and federal authorities inherent in this arrangement. To render its investment more secure, for example, the federal government acquired water rights under state laws, and later transferred them to the individual farmer after he had paid for the irrigation works and had assumed ownership of the distribution system. Until that time, however, water rights which the federal

tion Congress, 1903, 169–174. Bien's work was criticized in *Irrigation Age* (Jan. 1909), 24, 70–71.

[37] The Act, however, was influential in water law reform in other states. See, for example, *Reno Gazette*, Oct. 17, 1892 (FGN, Scrapbook #6).

[38] For an account of Mead's early work see Fred Bond and J. M. Wilson, "The Irrigation System of Wyoming," United States Department of Agriculture, Office of Experiment Stations, *Bulletin #96, Irrigation Laws of the Northwest Territories of Canada and of Wyoming*, 47–90.

[39] J. M. Wilson described the problems with which the Office dealt in *Proceedings, National Irrigation Congress, 1900*, 14–25. See also J. C. True to Gifford Pinchot, Dec. 10, 1913 (GP #1937).

government owned were subject to state law. By the terms of the Act of 1902, moreover, the federal government retained title to the reservoirs and large ditches, and agreed to continue their maintenance and operation forever. Under these circumstances, farmers owning water rights under state law would be at the mercy of federally operated reservoirs which stored state-controlled water. Conflicts between state and federal authorities arose frequently, especially when the Reclamation Service tried, by cutting off the water supply, to force settlers to meet their legal obligations to the federal government.

An even more troublesome problem confronted the Reclamation Service: water in one state often could most efficiently irrigate lands in another. Yet the transfer could rarely be accomplished. In Nevada, for example, Newlands and others waged an unsuccessful campaign to obtain water from Lake Tahoe in California to irrigate lands in the lower Truckee Valley in Nevada.[40] Newlands had hoped that the federal government could plan for full development of interstate streams by retaining the freedom to locate reservoirs and irrigable lands irrespective of state lines. Yet, the Reclamation Service met great resistance from local people who wanted to use the water in their own state and complained of federal interference with state rights.[41]

[40] This complicated but little known controversy over Lake Tahoe is described in letters in the Newlands collection over a period of fifteen years, and also in the papers of William Kent, California congressman, located in the Sterling Memorial Library, Yale University. For a version favorable to the California groups see Elizabeth T. Kent, *William Kent, Independent* (Private photo-offset printing, np, 1950), 318–324. The Nevada point of view can be traced in A. E. Cheney to Newlands, Dec. 26, 1900 (FGN, Letters); Newlands to President Wm. H. Taft, Aug. 14, 1909 (FGN, Irrigation—Tahoe-Truckee, 1909–1912); D. L. Noble to Newlands, July 11, 1913 (FGN, Scrapbook #13).

[41] In 1906 the National Irrigation Congress appointed a committee to examine this matter. It proposed that priorities across state lines be established by a federal administrative system corresponding in character to that needed for establishing and protecting rights within a state. See speech of Gov. George E. Chamberlain of Oregon before the Joint Conservation Congress, Dec. 8, 1908, as reported in *Conservation* (Jan. 1909), 15, 9–10.

Even before the Newlands bill became law, Western state leaders recognized the degree to which it would affect their water rights. In the summer of 1901, meeting at Cheyenne, Wyoming, several Western state engineers drew up a rival measure which provided that the federal Reclamation Fund be distributed through the state engineers who would construct projects approved by the Secretary of the Interior.[42] In this way state officials hoped to combine federal financing with a minimum of federal interference with local water law. Although Western congressmen considered this measure late in 1901, Newlands vigorously opposed it and it was laid aside.[43] Newlands, however, agreed to an amendment to his measure which emphasized that water from federal projects would be distributed and used under state law.[44] This concession calmed Western fears momentarily, but hardly solved the problem permanently. State-federal conflicts over water rights continued to plague the Reclamation Service. Federal officials, in fact, frequently advocated, though without much hope for success, that a federal water law supplant state statutes.

Speculation Interferes with Planning

The speculative predispositions of Westerners interfered with a program of efficient water development as much as did state water law. The problem of speculation, in fact, revealed the degree to which the petty land shark could paralyze large-scale water projects, whether carried out by private or public agencies. Earlier irrigation diversion works had been cheap, temporary, and simple to construct. Large reservoir and ditch systems, however, required detailed technical data, a secure water right, and long-term operation sufficient to repay the large capital investment. The planning and stability essential for such a program were difficult to foster in a frontier area of rising land values, quick profits, and rapid change. Not yet settled down to permanent and stable development, the West, even in the

[42] *The Tribune* (no further identification), nd (FGN, Scrapbook #12, p. 59).

[43] Newspaper clipping (no further identification), nd (FGN, Scrapbook #12, page 61); see also "Bartine MSS" in FGN, Irrigation, History of National, 2.

[44] Newlands to William F. Herrin, Feb. 5, 1902 (FGN, Letters).

early twentieth century, exhibited many attitudes which ran counter to the spirit of efficient planning. Most important among these was the desire and opportunity to speculate.

Federally owned lands had always offered a great opportunity for Americans from all walks of life to reap profits from rising values,[45] and Westerners proved no exception. Each settler hoped to make a capital gain. Obtaining lands cheaply through homestead entry, he sold them for a tidy sum to another who did not want to wait five years to prove up his claim. After passing through the hands of several owners, each of whom plucked his share of unearned increment, lands purchased originally at $1.25 an acre reached $500 or even $1000 an acre in areas where irrigation gave rise to fruit and vegetable farming.[46] Speculation increased investment costs for every purchaser interested in permanent development, and especially to farmers under irrigation projects. Most speculators had no intention of settling down to work their land as a continuing investment, seeking only to "cut out and get out" of the business of land ownership. They took a short-run attitude toward land ownership and land values as readily as they exploited natural resources without thought for future economic growth. Speculation was not so much the work of large corporations, as of fly-by-night sharks, most often operating on small capital. Only the larger firms could provide the long-term investment and stability essential for more rational development.

Both private and public irrigation promoters tried to restrain spiraling land values and prevent speculation from interfering with their plans. The fact that land titles came from the federal govern-

[45] Historians have identified the Western speculator with the large corporation, and have considered speculation as one phase of the struggle between the concentration of corporate wealth and the "people." Men of small means, however, speculated just as frequently as did men of large means. Speculation was significant less as an aspect of social and economic conflict, and more as a problem in economic development. It thwarted large corporate as well as federal enterprise.

[46] Frederick H. Newell discusses the effect of speculation on irrigation development in an article, "Irrigation Finance," MSS dated Feb. 28, 1913 in FN #6.

ment and water rights from the states, creating divided ownership of land and water, rendered every ditch and reservoir company an easy victim for the enterprising land shark.[47] Although the company owned the ditches and water rights, its success depended upon the willingness of farmers to take up land from the federal government, only to settlers who purchased water rights from the company for the water. Many settlers, however, did not intend to become permanent farmers; they obtained land, waited for values to rise, and sold out. Since ditch owners could not force a farmer to purchase a water right, they often received less revenue than anticipated and went bankrupt.[48] Such experiences gave rise to a provision in the Carey Act of 1894 that land in these new projects be sold only to settlers who purchased water rights from the company which constructed the works. The author of the Act, Senator Robert Carey of Wyoming, had faced difficulties with speculators in irrigation projects which he had promoted, and hoped to avoid similar problems in the new program. In the Reclamation Act of 1902 the provision that land ownership and water rights be combined afforded equal protection for federal projects.

These measures, however, did not restrain speculators who staked out claims on strategic reservoir, ditch, and farming sites before the Reclamation Service approved a project for the same area. That agency's attempt to keep its plans secret hardly succeeded when a Western community had boosted a project for years and speculators had entered the field early. Both private and public promoters tried to solve this problem by recommending that Congress authorize the Reclamation Service to withdraw from entry all land capable of being irrigated. Finding that settlers used the Desert Land Act, in particular, to acquire potentially valuable irrigation sites for speculation, they also demanded repeal of that Act. By 1908 complaints

[47] Many soon realized the disadvantages of this dual jurisdiction. "It is now realized," Elwood Mead, for example, wrote, "that the federal government should have asserted the same ownership over the public water that it did over the public land, and disposed of both together. Rights to streams could then have been acquired by some orderly and systematic administrative procedure." Mead, *Irrigation Institutions*, 62.

[48] *Ibid.*, 20–22.

from private groups persuaded the Commissioner of the General Land Office to recommend that Congress immediately withdraw from entry all remaining irrigable land in the West. Only in this way, he argued, could promoters thwart speculators and proceed with projects in an orderly manner.[49]

Irrigation and Forest Cover

A number of water supply problems brought the irrigation movement into close connection with forestry. In fact, the conservation movement of the Roosevelt administration grew out of a fusion of land and water policies which took place around the turn of the century. Many historians have not sufficiently emphasized the close connection between forestry and irrigation. Although he pays tribute to Newell and to the political support for forestry by the organized irrigation movement, Gifford Pinchot, in his major work on the Roosevelt Era, *Breaking New Ground*, minimizes the role of water development in the larger conservation movement of 1907–1908. Charles R. Van Hise wrote more accurately of the historical development: "It was seen by Mr. Pinchot and other scientists . . . that there is a close connection between the forests and waters. There was a strong public demand that our rivers maintain a uniform flow for water powers and for navigation. Therefore those primarily interested in forests and those interested in waters became associated in the conservation movement."[50]

Western irrigators pioneered in the theory that watershed vegetation directly affected their water supply. Forests, they argued, absorbed rainfall, retarded stream run-off, and increased the level of ground water; forests retarded snow melting in the early months of the year, reduced spring floods, and saved water for summer use when supplies ran low; forests retarded soil erosion and silting in

[49] U. S. Department of the Interior, *Report of the Commissioner of the General Land Office, 1908*, 14–15.

[50] Charles R. Van Hise, *The Conservation of Natural Resources in the United States* (New York, 1910), 5.

irrigation ditches and reservoirs.[51] Private power and water supply corporations, as well as municipal water departments, joined with irrigators in presenting these arguments.[52] They opposed commercial use of the watersheds; they fought to prevent lumbering in the forests and grazing on the mountain ranges. They centered their fire especially on sheep, which cropped vegetation close to the ground and, they argued, vastly accelerated erosion.[53]

Western irrigators played a major role in establishing the national forests and in defending them from attack. The primary intent of Congress in setting aside forest reserves in fact was watershed protection. The chairman of the conference committee from which the Act came, for example, explained: "We have made a provision in this bill authorizing the President of the United States whenever in his judgment he deems proper to do so, to make a reservation of the timber lands, principally applying in the watersheds of the West, so that the water supply in the country may be preserved. . . ."[54] Throughout the nineties irrigation groups petitioned, often successfully, that the president reserve particular watersheds as national forests to protect them from commercial use. At the same time, the National Irrigation Congress supported the federal forestry program. Irrigators constantly sought to improve protection of the forests from fire and timber depredations, to withdraw them from all commercial use, and to prevent timber cutting and grazing within them.[55]

[51] For these views see Colorado Experiment Station, Ft. Collins, Colorado, *Weekly Bulletin #28*, described in *The Forester* (April 1899), 5, 85; L. G. Carpenter, *Forests and Snow, Colorado Agricultural Experiment Station Bulletin #55* (Ft. Collins, 1901); William Wallace Pardee to Edward A. Bowers, March 27, 1893 (GP #1674); James D. Schuyler, "The Influence of Forests Upon Storage Reservoirs," *The Forester* (Dec. 1899), 5, 285–288.

[52] Adolph Wood to Edward A. Bowers, Mar. 23, 1893 (GP #1674).

[53] These attitudes are expressed, for example, in *The Forester* (May 1898), 4, 96, and by Representative Loud of California in 56th Congress, 1st Session, *Congressional Record*, 1446, 5522.

[54] 51st Congress, 2nd Session, *Congressional Record*, 3547.

[55] For grazing problems on the forest reserves in the 1890's see John Ise, *The United States Forest Policy* (New Haven, 1920), 121, and U. S. Department of the Interior, *Annual Report of the Commissioner of the General Land Office, 1899*, 108–110.

These views prevailed especially in California where steep slopes and torrential rainfall created acute flood and erosion problems. Urbanites concerned with water supplies and irrigators led in demanding public action to protect forest cover. Early in the 1890's, for example, President Benjamin Harrison, responding to petitions from southern California groups, created the San Bernardino National Forest. The leader of this particular movement was General Adolph Wood, president of the Arrowhead Reservoir Company, a corporation engaged in storing water for power, irrigation, and domestic supply. By 1899 Wood and others of like mind had organized the California Water and Forest Association, which agitated for more adequate state laws to protect forests, encouraged tree planting on denuded watersheds, and advocated state cooperation with the U. S. Geological Survey to measure water resources. When the state legislature failed to grant an appropriation to finance water investigations, the Association itself provided the funds to match federal contributions for the Survey's work. The Association attacked grazing in the forests, argued that livestock should be excluded, and persuaded the Department of the Interior to institute proceedings against forest trespassers. Responding to petitions from many groups throughout the state, Presidents Cleveland, McKinley, and Roosevelt greatly extended California reserves. When in 1907 Congress prohibited the president from establishing more reserves in many Western states, it expressly excluded California from the law.[56]

Irrigators in other states displayed similar concern over forest cover. Colorado State Engineer John Field described the benefits to irrigators from forested watersheds and encouraged and applauded the efforts of the Colorado Forestry Association to extend the federal reserves in that state. Hydrographers at the Colorado State College of Agriculture at Fort Collins wrote of the close connection between

[56] The California movement can be traced in *California Illustrated Magazine* (Nov. 1892), 2, 792–807; (Nov. 1893), 4, 841–850; in *The Forester* (Jan. 1899), 5, 18, (Feb. 1899), 5, 38, (Mar. 1899), 5, 65, (June 1899), 5, 136–137, (July 1900), 6, 170–171, (Oct. 1901), 7, 244–250; and in *Forestry and Irrigation* (Nov. 1904), 10, 521.

forests and stream flow and of the detrimental results of stock-grazing on the headwaters of the South Platte River. In Arizona, irrigators on the Gila River near Phoenix opposed the commercial use of the watersheds above them. After the territorial legislature memoralized the federal government to protect the headwaters of the Gila, the Department of the Interior established three forest reserves there. For a time Arizona irrigators were able to exclude grazing entirely from these reserves, but after protests from stock-men and intervention by Gifford Pinchot, the Department permitted a limited amount of grazing.[57]

Engineers and scientists in the new water and forest departments of the federal government shared the view that watershed cover improved water supplies. The Division of Forestry carried out numerous studies of the effect of forest cover on stream flow. James W. Toumey of the Division undertook the first such investigation in southern California, and reported that run-off from forested areas was far below that from non-forested. Hydrographers in the Geological Survey, with even wider experience, gave the theory vigorous support. The national forests, so they argued in governmental reports and public speeches, contributed immeasurably to the conservation of water resources.[58]

These common views brought the forest and irrigation associations into close contact; they supported each others' programs before Congress and the country at large. During the 1890's, while Frederick H. Newell was undertaking hydrographic investigations and agitating for a national irrigation law, he also served as secretary of the American Forestry Association, which fought to extend the national forests and to adopt a sound national forest management

[57] *The Forester* (Feb. 1899), 5, 41, (Mar. 1899), 5, 65; C. S. Crandell, "Reproduction of Trees and Range Cattle," in *Ibid.* (July 1901), 7, 170–174; U. S. Department of the Interior, *Annual Report of the Commissioner of the General Land Office, 1899*, 98; Pinchot, *Breaking New Ground*, 177–181.

[58] Bernhard E. Fernow, *Relation of Forests to Water Supplies, United States Department of Agriculture, Division of Forestry Bulletin 7* (Washington, 1893), 123–170; James W. Toumey, "Relation of Forests to Stream Flow," *United States Department of Agriculture Yearbook, 1903* (Washington, 1904), 279–388; *New York Tribune*, Jan. 11, 1904 (FN #17, Clippings, v. 4).

program.[59] George Maxwell and the National Irrigation Association supported the forest movement as actively as they campaigned for federal irrigation, with the motto, "Save the Forests, Store the Floods, Make Homes on the Land."[60] Forestry enthusiasts, in turn, promoted federal irrigation. In 1901 the American Forestry Association changed the name of its official magazine to *Forestry and Irrigation*, and proceeded to publicize the irrigation movement. Gifford Pinchot, chief of the Bureau of Forestry, attended sessions of the National Irrigation Congress, spoke frequently on the beneficial influence of forests on stream flow, and joined with Newell in pushing the Newlands reclamation measure.[61]

The close association of Newell and Pinchot in forest and irrigation matters illustrated a common attitude toward resource development then emerging in the federal government. Members of the Bureau of Forestry and the Geological Survey, in particular, developed a similar outlook. They became personal friends, came together frequently in meetings of scientific societies in Washington, and gave each other mutual encouragement and political support. At first preoccupied with a federal water resource program, they next turned their attention to forestry and then to an increasing variety of problems in which they could apply their interest in rational and efficient development. They became the nucleus of a group of federal scientists and technicians whose search for greater efficiency in economic growth gradually committed the administration of Theodore Roosevelt to a wide program of natural resource conservation.

[59] Newell was corresponding secretary of the Association at least as early as 1897, and resigned when he was placed in charge of federal irrigation work. As early as 1889 he had been on the legislative committee of the Association. See Edward A. Bowers to H. H. Chapman, Nov. 18, 1916 (GP #1674).

[60] Through Maxwell's influence, business organizations also backed the national forest movement. See resolutions of the National Board of Trade, the National Business League, and the National Association of Manufacturers in *The Forester* (Feb. 1901), 7, 47–48; (Nov. 1900), 6, 273; *Forestry and Irrigation* (June 1904), 10, 243.

[61] For several years Pinchot was chairman of the forestry section of the National Irrigation Congress, and Pinchot, rather than Newell, was President Roosevelt's personal representative at those Congresses. See Pinchot, *Breaking New Ground*, 189–191.

Chapter III

Woodman, Spare that Tree

While federal forestry officials joined hydrographers in promoting irrigation, they also campaigned for a more rational and efficient use of timber resources. The lumber industry provided a dramatic example of resource exploitation in the last half of the nineteenth century. Low prices for forest land and the meager amount of capital required encouraged many enterprising Americans to enter the field. The industry was highly mobile; lumbermen rapidly cut and abandoned lands, and moved on to new areas. Facing the uncertainties of a high degree of competition, timber operators could not, with profit, utilize fully their resources. Few cut more than the most valuable trees; they left second-grade timber as waste. Lumbermen showed little concern for fire and disease damage, the destruction of trees during logging, or provisions for adequate reproduction. By the latter part of the nineteenth century the obvious waste resulting from these practices prompted a few leaders in both public and private forestry to demand that the country give more attention to a program of scientific management.

Toward a More Stable Forest Industry

As early as 1875 some of the few Americans concerned with this problem organized the American Forestry Association. Composed primarily of botanists, landscape gardeners, and estate owners, this group emphasized arboriculture, an aesthetic appreciation of forests, and the study of individual trees. It displayed more concern for saving trees from destruction than for efficient timber management.

The Association agitated for state laws to encourage tree planting, helped to establish Arbor Day, promoted the plan for national forest reserves, and was responsible for the development of many parks. It did not deal with forestry as an economic problem, or cooperate with timbermen to establish less wasteful practices. The American Forestry Association, during its early years, played a negative role of drawing public attention to timber destruction, but it hardly blazed the trail toward the positive goal of more efficient forestry.

During the 1890's the organized forestry movement in the United States shifted its emphasis from saving trees from destruction to promoting sustained-yield forest management. According to this viewpoint, to provide a continuous supply of timber for the future, annual cutting should never exceed annual growth, and lumbermen should utilize waste materials and reduce fire and disease damage. This concept of sustained-yield, or planned, long-range timber management was hardly in tune with the fluid, rapidly shifting economy of the latter half of the nineteenth century. Most lumbermen looked upon it as an impractical dream. At the same time, the prevailing forest reformers preferred to prohibit all timber cutting. Scientific forestry, however, steadily became more popular as a growing group of professional foresters gradually overcame both these hurdles.

Gifford Pinchot, a young German-trained forester of considerable means, ability, and enthusiasm, assumed leadership of this new movement.[1] Born in Connecticut in 1865, Pinchot graduated from Yale in 1889. His father encouraged him to take up the profession of scientific forestry, but forest leaders with whom young Pinchot discussed the matter, disagreed. Sustained-yield management, they argued, was not yet profitable in the United States; therefore, the nation could not yet support a separate forestry profession. Pinchot was more optimistic. He spent several years in France and Germany, studying the most advanced forest practices, returned to the United States, and in 1892 took charge of timber management on the Biltmore estate in North Carolina. A year's experience with selective

[1] Pinchot, *Breaking New Ground*, 1–78, contains the best summary of the early years of the new forest movement.

logging, brush-burning, and fire control convinced the youthful enthusiast that he was on the right track, and should expand his activities. He published the results of his work at Biltmore, served as a professional consultant to private forest owners, drew up working plans for private forest management, and wrote and lectured widely on the practicability of scientific forestry.

In 1898 Pinchot became Chief of the Division of Forestry, and, after 1900, of the new Bureau of Forestry in the Department of Agriculture. Organized in 1879, this Division had been directed by Bernhard Fernow, a German-born, German-trained forester.[2] Fernow's views differed radically from those of Pinchot. Convinced that neither the public nor the forest industry would yet support scientific management, Fernow believed that the Division of Forestry should merely dispense information and technical advice to those who sought it, and not promote sustained-yield practices. Confining its investigations to the study of individual trees, the Division did not experiment with new techniques which might improve forest management.

When Pinchot became Chief, he radically changed the Bureau's emphasis. He set out to educate the public and the private forest industry about scientific forest management. For example, to help private owners draw up management plans, he offered the services of federal foresters. In reply, timber companies swamped the Bureau with requests for aid. Some of the largest timberland owners in the country, among them the Kirby Lumber Company of Texas, the Northern Pacific Railroad, and the Weyerhaeuser Lumber Company in the Pacific Northwest, took up the offer. By 1905 owners of some three million acres had applied for assistance. Bureau officials had already investigated most of this area, and had persuaded private owners to place 177,000 acres under better management. The new Chief also undertook investigations more pertinent to scientific forestry. To prove to lumbermen in dollars and cents that fire protection would pay, he conducted detailed inquiries into

[2] A recent biography of Fernow is Andrew Denny Rodgers, III, *Bernhard Eduard Fernow: A Story of North American Forestry* (Princeton, 1951).

forest fire destruction. And a section of the Bureau studied tree planting and advised forest owners about reforestation problems.[3]

Through sheer force of personality and conviction, Pinchot drew many enthusiasts into different phases of the forest movement. His vigor and drive captured the interest and loyalty of a number of young men, such as the future Chief Forester, Henry S. Graves. His detailed knowledge of forestry and his concern for making forestry pay attracted the friendship of many practical lumbermen. One of these, Eugene Bruce, an Adirondack logger, became a mainstay of the future U. S. Forest Service.[4] In 1898, partly as a result of Pinchot's urging, the New York legislature established the Cornell Forest School to train scientific foresters, and in 1900 his father, James Pinchot, contributed an initial endowment to launch a forest school at Yale. In 1901, foresters with scientific training organized the Society of American Foresters. Playing an increasingly active role in the American Forestry Association, the same men changed that group's emphasis from arboriculture to scientific forestry.[5]

The businesslike approach of Pinchot and his co-workers persuaded many in the forest industry to join in the new movement. Lumbermen took part in the work of the American Forestry Association.[6] For three years, when Congress failed to appropriate the money, they provided funds to pay a full-time clerk in the office of the Bureau of Forestry.[7] They raised $125,000 to endow a chair at the Yale Forest School.[8] They opened their facilities and timber-

[3] The most concise summary of these developments is in Pinchot, *Breaking New Ground*, 133–161.

[4] Bruce has left an invaluable account of his work in the Forest Service and his relationship to Pinchot. See Eugene S. Bruce, "Recollections of Pioneer Forestry," MSS in GP #1714.

[5] Pinchot, *Breaking New Ground*, 150–152.

[6] In 1905 George K. Smith, President of the National Lumber Manufacturers' Association, became a director of the Association. By 1909 the Association had an advisory board including representatives of nine lumbermen's organizations.

[7] Testimony of J. B. White before the "Trade Committee," Feb. 16, 1914, in GP #1668.

[8] *Forestry and Irrigation* (July 1910), 16, 381.

lands to forestry students for practical field training. They provided Pinchot with crucial political support in warding off congressional attacks on the Forest Service.[9] Many in the industry scorned "conservative" lumbering and federal forestry, but a growing body of influential leaders gradually brought private and public officials more closely together.

A common concern for future timber supplies helped to bring about this *rapprochement*. Lumbermen, for example, could no longer move on to virgin areas when they had exhausted their available timber. They now became more interested in using existing supplies more efficiently. They took up fire protection, utilization of low-grade wood, and measures to guarantee reproduction. Wood-consuming industries shared these concerns. Vehicle, box, and furniture manufacturers, for example, secured satisfactory supplies with increasing difficulty. If they could not obtain more reliable sources of lumber, they would be forced to curtail operations.[10] Hardwood users became especially interested in the proposal to set aside the Appalachian mountain range as a sustained-yield hardwood area. They became active in the American Forestry Association and swung their trade organizations behind the U. S. Forest Service.[11]

As a result of this common concern, private and public leaders joined in developing a fire control program. The success of fire control measures in the federal forest reserves, as well as studies published by the Bureau of Forestry, convinced many lumbermen that such a program was profitable. This point of view advanced most rapidly among private companies in the Pacific Northwest. As a result of disastrous fires there in the summer of 1902, private

[9] *American Conservation* (March 1911), 1, 39.

[10] William L. Hall, *The Waning Hardwood Supply and the Appalachian Forests, U. S. Department of Agriculture, Forest Service Circular #116.*

[11] Among the members of the advisory board of the American Forestry Association were representatives of the Tight Barrel Stave Manufacturers' Association, the National Association of Box Manufacturers, the Carriage Builders' National Association, and the National Slack Cooperage Manufacturers' Association. See *Forestry and Irrigation* (Jan. 1909), 15, 1; see also memorandum, Harry A. Slattery to Pinchot, July 23, 1913 (GP #1828).

owners organized fire protective associations, promoted state legislation to aid fire control, and financed fire patrol and fire suppression activities on their own lands.[12] In 1909 these fire associations combined to form the Western Forestry and Conservation Association. Its manager, E. T. Allen, a former official of the U. S. Forest Service, led the Association in meeting the problems of that area and in awakening lumbermen elsewhere to the financial advantages of fire protection.[13] Both public and private groups sponsored the Weeks Act of 1911 which provided federal financial aid to any state that would take up a program to protect from fire the timberlands at the headwaters of navigable streams. Under the leadership of William B. Greeley, the Forest Service administered this program and stimulated an expansion of fire prevention activities throughout the nation.[14]

Public and private leaders also cooperated in promoting research in new uses for waste materials and new processes for manufacturing forest products. Fernow had confined his research to "timber physics" and in 1896 the Secretary of Agriculture had discontinued even this as "not germane to the subject of the Division." In 1901 Pinchot organized a forest products section in the Bureau which conducted investigations in the chemistry of maple sugar, tree diseases, and methods of extracting turpentine, all of which yielded information of commercial value. In 1909 the Forest Service transferred this work to a central forest products laboratory at Madison, Wisconsin. The University of Wisconsin provided a building, and the Forest Service and private lumbermen purchased the necessary equipment.[15]

Federal foresters and private lumbermen also developed a common interest in maintaining timber prices. The Forest Service, well

[12] William B. Greeley, *Forests and Men* (New York, 1951), 19–20.

[13] Accounts of these activities are in *American Forestry* (April 1910), 16, 256–257; (May 1910), 16, 357; *Proceedings, National Conservation Congress*, 1911, 114–116; Greeley, *Forests and Men*, 21.

[14] *Ibid.*, 24–29. Another important group was the Northern Forest Fire Protective Association in the Lake States; see *American Forestry* (Dec. 1910), 16, 750.

[15] Pinchot, *Breaking New Ground*, 306–313.

aware that scientific timber management meant higher costs, shared the concern of lumbermen over low prices for their products.[16] As Pinchot remarked: "We have got to make the public see that cheap lumber is not good for forest conservation."[17] Few industry leaders, of course, used similar reasoning. Most wished to raise prices merely to increase their profits and not necessarily to inaugurate sustained-yield management. But those who were most concerned about conservation echoed Pinchot's viewpoint that more expensive timber practices depended entirely on adequate profit margins.

Professional foresters supported the lumber industry in a number of efforts to raise prices, for example, through a higher tariff. Officials of the Forest Service could not afford to ignore the effect of foreign competition, either in lower profits or in greater waste of low-grade timber. A tariff would help maintain higher prices and render conservation profitable. In March 1909 Pinchot wrote to Sereno E. Payne, chairman of the House Ways and Means Committee, recommending a higher duty on lumber.[18] Higher prices, he argued, would encourage use of low-grade timber. This was Pinchot's own personal viewpoint. He refused to make an official statement on the tariff or to commit himself when asked directly. Fearing that in a pinch lumbermen would place more emphasis on profits than on long-range management, he declined to identify the Forest Service too closely with the industry.

The tariff question illustrated the tenuous relationship between the Forest Service and the forest industry. Both groups faced common problems. Both knew that large-scale lumber operations could increase efficiency and reduce unit costs so as to absorb the added expenses of sustained-yield management. With their large capital investments, lumber corporations had a sufficient stake in the future to worry about prospective timber supplies. Yet Pinchot also feared that the industry's desire to maximize profits might outweigh its tentative acceptance of sustained-yield management. Therefore, although common problems drew the Forest Service closer to

[16] Philip P. Wells to J. B. White, May 13, 1914 (GP #1827).
[17] *Proceedings, National Irrigation Congress, 1909*, 514.
[18] Pinchot to Sereno E. Payne, Mar. 10, 1909 (GP #1819).

private industry, Pinchot hesitated to establish ties with lumbermen which he could not control or which could not be severed quickly.

This paradoxical relation became clear when the Southern yellow pine industry attempted to form a corporation sufficiently large and powerful to control production and establish improved timber management. During the panic of 1907 the yellow pine industry suffered a severe depression. To solve their problems, which they attributed to overproduction, yellow pine manufacturers agreed to form a corporation to purchase 60 per cent of the stumpage in Missouri, Arkansas, Louisiana, Texas, Mississippi, and Alabama. The new firm would not only control production, but would achieve economies through savings in railroad construction, specialization of mills, and coordinated executive and sales departments. The industry feared, however, that the public would bitterly oppose such a combination. To forestall this reaction, industry leaders proposed to the Forest Service an agreement whereby the lumbermen would carry out improved forest practices if federal officials would approve the new corporation.[19]

Pinchot received the plan with great favor. Combination, he argued, would stabilize the industry, increase profits, and encourage sustained-yield forestry. He did not have faith, however, that the industry would take up scientific management voluntarily. As a price for his approval he insisted that the new corporation's officials commit themselves in writing to carry out specific conservation practices.[20] He proposed that the two groups sign a contract in which the lumbermen would agree to pay the expenses of management and the Forest Service would perform the work. Rejecting this alternative, the manufacturers proceeded to form their combination, but were soon thwarted in the courts by Attorney-General Herbert S.

[19] See MSS, "Present Condition of the Yellow Pine Industry" (GP #97); Nelson W. McLeod to Pinchot, Feb. 24, 1908 (GP #97); *Boston Transcript*, July 9, 1908 (GP #1668).

[20] Pinchot to Nelson W. McLeod, March 4, 1908 (GP #97); Pinchot to William T. Cox, March 27, 1908 (GP #100).

Hadley of Missouri.[21] A few years later, when Congress was drawing up new anti-trust legislation in 1913 and 1914, the lumbermen sought exemption from proposed restrictions by arguing that conservation required combination.[22] Henry S. Graves, the new Chief Forester, refused to support this plea unless the lumbermen would agree to public control, and they would not.[23] Yet forest officials equally criticized those who attacked industry-wide cooperation on principle. President Wilson and his advisers, they complained, seemed to dread "*any* toleration of combinations to restrict output, no matter how conditioned."[24]

Even though private and public forest leaders differed about federal regulation of timber practices, the new scientific forestry spirit clearly appealed to both groups. Both wished to promote a stable and permanent lumber industry, utilizing the most advanced technology. It was hardly surprising, therefore, that timber corporations should find much that was attractive in the federal forestry program. Scientific forestry was a characteristic example of the new technological age which emphasized large-scale, long-term planning and management in both private and public affairs.

The Federal Forest Program

Private forest owners, although they indicated a great interest in the new approach to forestry, did not translate interest into practice rapidly enough to please Pinchot.[25] The national forests, he came to feel, offered a far better opportunity to demonstrate scientific forestry, and he soon placed primary emphasis upon federal forest

[21] Philip P. Wells, "Cooperation in Commercial Business and in Control of Private Forest Lands," MSS in GP #1671. See also "Proposed Contract between the Forest Service and the Southern Lumbermen's Organization," MSS in GP #1668.

[22] Testimony of J. B. White before the "Trade Committee," Feb. 16, 1914, in GP #1668.

[23] Philip P. Wells to Pinchot, Feb. 23, 1914 (GP #1668).

[24] Wells to J. B. White, March 4, 1914 (GP #1668).

[25] Greeley, *Forests and Men*, 102–103. By 1920 Pinchot had stopped trying to cooperate with the lumber industry and was seeking federal regulation of timber cutting on private lands.

management. The vast stands of virgin timber on the public lands formerly had passed rapidly to private ownership. Throughout the 1880's, the American Forestry Association and the American Association for the Advancement of Science advocated that the federal government change this policy and hold Western timberlands permanently as public reservations.[26] Joined by irrigators, and with the aid of Secretary of the Interior John Noble and President Benjamin Harrison, these groups secured an amendment to the General Land Law Revision Act of 1891 which granted the president authority to create forest reserves by proclamation.[27]

The Act of 1891 merely established reserves; it did not provide for their management. On this question there was wide disagreement. Those who looked upon the forests as preserves which should remain untouched to protect watersheds or serve as areas of natural beauty wanted a program in which the Army would patrol the forests to exclude timber thieves, stockmen, and other interlopers. To others, such as Gifford Pinchot, management involved much more: the development of a trained forestry force to control fires, tackle disease problems, and supervise cutting and sales, as well as to maintain the "integrity" of the forests.[28] In passing the Act of 1891, Congress had reflected the interests of the first group—that the forests be reserved from commercial use—and the Secretary of the Interior administered the Act in accordance with this spirit.[29]

A new law, the Forest Management Act of 1897, granted the Secretary of the Interior power "to regulate the occupancy and use" of the reserves, and provided the opening wedge for the rational development which Pinchot preferred. That Act, although it did not explicitly recognize grazing as a legitimate use of the forests, did

[26] John Ise, *U. S. Forest Policy*, 110–118.

[27] Edward A. Bowers to H. H. Chapman, Nov. 24, 1916 (GP #1674).

[28] An account of problems leading up to management legislation is in Pinchot, *Breaking New Ground*, 86–104.

[29] The General Land Office and the Department of the Interior emphasized that the forests were to be used primarily to protect water and timber supplies, and that grazing was only secondary. They acted in accordance with these objectives. See Peffer, *Public Domain*, 72–73.

not expressly forbid it, thereby permitting the administering agency to decide the specific meaning of the general grant of regulatory power. Mining companies, on the other hand, had thwarted the efforts of irrigators and game and park preservationists to prohibit lumbering in the reserves, and had forced into the Act a provision that dead and down timber could be cut and sold. Moreover, the Act gave to miners the right to stake out claims and obtain patents within the forests as readily as on other Western lands. Although the Act of 1897 did not specifically grant full commercial use, it failed to exclude it and firmly established the Secretary's authority over the reserves. Federal forest policy could now turn in a new direction. The Act paved the way for federal officials in the future to permit grazing, commercial lumbering, and hydroelectric power generation within the forests, and to establish the national forest program clearly as one most concerned with rational development.

The Secretary of the Interior placed the administration of the reserves in a Division of the General Land Office, and during the years between 1897 and 1905 took steps toward the development of a forest service. The Division instituted a fire prevention and fire suppression program, which greatly reduced losses. It inaugurated a system of timber sales and a tree planting program; it drew up a timber management plan which, however, it failed to implement. Pressure from Western stockmen persuaded the Department to permit some grazing, though under strict regulation. In most of this work the Secretary sought advice from the Bureau of Forestry which conducted field investigations and recommended management policies. Insufficient appropriations handicapped the new program, but by 1903 the Secretary had succeeded in raising the original annual grant of $75,000 to $300,000.[30]

The General Land Office carried out its new task only with great difficulty. Most of its officials were law clerks, trained in the legal details of land disposal, but thoroughly unfamiliar with forestry or the West. Yet the Department required them to draw up and

[30] These developments can be traced in the successive annual reports of the Secretary of the Interior and the Commissioner of the General Land Office, 1897–1905.

administer forest rules. Trained as lawyers, they had no large views of the possibilities of forest management, but adhered strictly to narrow interpretations of law and emphasized formal procedures rather than results. The custom of political appointments to the General Land Office hampered the selection of technicians. Politicians considered the position of forest supervisor as a patronage plum, for example, and bitterly criticized the General Land Office when it selected trained men for the post. The scientists and engineers in the Bureau of Forestry and the Geological Survey constantly ridiculed the Land Office "foresters," charging them with political favoritism, inefficiency, and ineptness. Deeply resenting these criticisms, the "foresters" fought back against every attempt to transfer their duties to any other federal agency.[31]

A few officials of the General Land Office tried to overcome these difficulties. Secretary of the Interior Ethan A. Hitchcock devised an arrangement with Pinchot whereby the Department of the Interior would patrol the reserves, while the Bureau of Forestry would direct investigations, make decisions on all technical forest matters, and execute plans. However, Commissioner Binger Hermann and his subordinates in the General Land Office successfully opposed this innovation. Hermann insisted particularly on retaining control over grazing and timber applications, a condition to which Pinchot refused to agree. Shifting his course, Hitchcock then created in the Department a new Division of Forestry to manage the reserves. Responding to Pinchot's advice, the Secretary placed a trained forester, Filibert Roth, in charge of the Division, and assigned to it men from the Bureau of Forestry. But the Land Office routine, its overly legalistic bent, and its practice of political appointments interfered too much with Roth's forest management program, and after two years he resigned.[32]

As early as 1898 Pinchot became convinced that the General Land Office could not administer the reserves effectively; the events of the next few years only confirmed this belief. The Office, he argued,

[31] Pinchot, *Breaking New Ground*, 161–172.

[32] *Ibid.*, 192–197; Roosevelt to Pinchot, Oct. 18, 1901 (TR); Roosevelt to Hitchcock, Oct. 19, 1901 (TR).

hopelessly involved in a maze of political appointments, legalistic routine, and personal favoritism, was not really interested in effective forest management. No one could cut through these entrenched inefficiencies. To solve the problem Congress should transfer the program to an entirely new agency, the Bureau of Forestry. Forest reserve management, Pinchot argued, was a technical task which professionally trained men could best perform. While few employees of the General Land Office had ever seen a forest reserve, officials in the Bureau of Forestry were trained for the work and actually knew the West firsthand. Moreover, as a biological science, forestry logically fell into the work of the Department of Agriculture.

For seven years, between 1898 and 1905, Pinchot worked for the transfer. He cajoled congressmen, sought the aid of influential railroad, mining, and lumber corporations, and called a special Forest Congress in Washington to arouse public opinion.[33] With the aid of George Maxwell, he persuaded those behind the irrigation movement, such as the National Board of Trade, to support the transfer.[34] The American Forestry Association took similar action. Pinchot's first attempt to secure such a measure, in the winter of 1901–02, met crucial and effective resistance from the Republican congressional leaders. Since Secretary of the Interior Hitchcock backed the transfer, President McKinley gave it his personal attention, but though the President privately favored the move, he refused to push it in the face of Republican opposition. Theodore Roosevelt, of an entirely different frame of mind, supported the proposal during the campaign of 1900 and included a plea for it in his first message to Congress in December 1901.

Pinchot obtained influential support from Western congressmen by convincing them that his administration would benefit the West more than did the policies of the Department of the Interior. As his most persuasive argument he advocated that the reserves be opened more fully to commercial use. Secretary Hitchcock's preference that

[33] Pinchot to Mrs. Phebea Hearst, July 15, 1904 (GP #83); Pinchot to Chauncey M. Depew, Dec. 15, 1904 (GP #89); telegram Pinchot to Hermann von Schrenk (undated) (GP #83).

[34] *Forestry and Irrigation* (Feb. 1903), 9, 101.

commercial use be minimized had created considerable Western opposition. Such compromises as he had been willing to make, the limited grazing program, for example, satisfied no one. While stockmen demanded that grazing be extended, those who looked upon the forests as public parks and watershed areas argued that it be forbidden entirely.

In support of greater commercial use of the forests, Pinchot played a crucial role in preventing game organizations from transforming the reserves into parks and game preserves, a possibility which the stockmen greatly feared. Eastern game organizations presented a variety of proposals to Congress to carve game preserves from the forests. When their petitions to convert all the forests into game areas met vigorous opposition from the West, they sought to reserve designated tracts from commercial use as breeding areas. This, too, Western congressmen thwarted.[35] The same groups then tried to persuade the Secretary of the Interior to recommend that lands in the Wyoming forest reserve be transferred to Yellowstone National Park. Wyoming stockmen, fearing the certain loss of grazing lands, opposed this proposal as well.[36] Quietly, though effectively, Pinchot used his influence against all of these measures. The Boone and Crockett Club, for example, which Roosevelt helped to found and of which Pinchot was a member, sponsored the bill to reserve areas in the forests for game breeding. Yet Roosevelt himself declined to support the bill and Pinchot discouraged the Club from pushing it in Congress.[37]

Pinchot also intervened to open the forest ranges to more livestock. When the Secretary of the Interior prohibited grazing in Arizona reserves in 1900, the Chief Forester inspected the area on

[35] *Forestry and Irrigation* (May 1902), 8, 218; 57th Congress, 1st Session, *House Report 968*, 6–7. The major opposition to this move came from sheep and cattle growers. See William T. Hornaday, *Thirty Years War for Wildlife* (New York, 1931), 213.

[36] *Proceedings, American National Livestock Association, 1904*, 148.

[37] Roosevelt to Austin Wadsworth, May 20, 1902 (TR); C. G. LaFarge to Pinchot, Jan. 16, 1903, Jan. 21, 1903 (GP #79); Pinchot to LaFarge, Jan. 31, 1903 (GP #79); LaFarge to Pinchot, Dec. 11, 1903 (GP #83); Pinchot to LaFarge, Dec. 14, 1903 (GP #83).

the ground, decided that limited grazing would not damage water supplies, and persuaded the Secretary to admit cattle and sheep.[38] Western stockmen appreciated these on-the-spot investigations of actual forage conditions. As early as 1901 they openly approved Pinchot's approach to the grazing problem, and contrasted it with past methods used by the Department of the Interior.[39] In its conventions of 1901 and succeeding years, the American National Livestock Association passed resolutions expressing confidence in Pinchot's views and recommending transfer of the reserves to the Department of Agriculture.[40] Two of the men who later were to lead the West in bitter opposition to the Roosevelt administration's public land policies—Dr. J. M. Wilson of the Wyoming State Experiment Station, and Colorado's future governor, Elias M. Ammons—expressed hearty approval of the transfer. At the 1904 Livestock Convention, for example, Ammons recommended that the reserves be turned over to "our friend, Mr. Pinchot. . . ."[41] In 1904 these and other spokesmen for the Western livestock industry were so friendly toward Pinchot that, on returning to Washington from the Convention, he could write, "Everything goes well and the transfer looks promising."[42]

Pinchot's opposition to "preservationists" and his support of grazing interests did not arise merely from his search for political backing for the transfer. These attitudes revealed his basic view that the reserves should be developed for commercial use rather than preserved from it. During the first years of his contact with the forests, in fact, Pinchot felt that his major problem was to restrain the influence of those who wished to leave them in their natural state, untouched by lumberman or stockman. At every opportunity he stressed the utilitarian value of the forests. "The object of our forest policy," he declared in March 1903 to the Society of American

[38] Pinchot, *Breaking New Ground*, 177–182.

[39] *Proceedings, American National Livestock Association, 1901*, 23, 145, 504–506.

[40] *Ibid.*, *1901*, 363; *1904*, 192; *1905*, 271.

[41] *Ibid.*, *1904*, 299–301.

[42] Pinchot to A. A. Anderson, Aug. 8, 1904 (GP #82).

Foresters, "is not to preserve the forests because they are beautiful . . . or because they are refuges for the wild creatures of the wilderness . . . but . . . the making of prosperous homes. . . . Every other consideration comes as secondary."[43] In an article prepared for *Century* magazine Pinchot elaborated on his views as to the objectives of federal policy, the administration's need for political support from forest users, and the obstacles placed in his path by naturalists, park enthusiasts, and wildlife groups. He wrote:

> The program of forestry stands upon practical conceptions. . . . Use must be the test by which the forester tries himself, for by it his work will inevitably be tried. . . . The test of utility has given the forest movement and the forest policy alike new strength and new acceptance. The misunderstanding of their objects and uses which has always been the chief local obstacle in the making of forest reserves necessarily yields before the argument of use, which implies also that no lands will be permanently reserves which can serve the people better in any other way. Forest reserves were never so popular as they are today, because they were never so well understood. For this result the President's Western trip last spring, during which he constantly advocated forest preservation for economic reasons, is largely responsible.[44]

Transfer measures met stiff opposition from the General Land Office. Although his superior, Secretary Hitchcock, supported the move, Binger Hermann, Commissioner of the General Land Office, opposed it because he wished to restrict grazing and to retain control of forest administration appointments. The Division of Forestry in the Interior Department also attacked the measure on the grounds that to transfer forest management to another department and leave land title matters in the General Land Office would create jurisdictional conflicts.[45] Pinchot minimized this possibility, but precisely such differences were to play an important role in the famous Pinchot-Ballinger controversy in 1909–10. Opposition from the

[43] Pinchot to Robert U. Johnson, Mar. 27, 1904, enclosing a copy of this address (GP #89).

[44] See MSS of this article in Pinchot to Johnson, Mar. 27, 1904 (GP #89).

[45] U. S. Department of the Interior, *Annual Report of the Secretary, 1901*, 56. Capt. J. B. Satterlee to Pinchot, May 30, 1907 (GP #97).

Land Office relaxed when William A. Richards, favorable to the transfer, succeeded Hermann in July 1903.[46]

To guide the bill through Congress, Pinchot at first relied on John F. Lacey of Iowa, chairman of the House Public Lands Committee and a staunch defender of the national forests. Lacey could not combat effectively the opposition of "Uncle Joe" Cannon, Chairman of the House Appropriations Committee, who attracted to his side a strong contingent of Eastern Republicans by arguing that the transfer would increase federal expenditures.[47] Pinchot's opportunity came when Frank Mondell, Republican representative from Wyoming and an archenemy of the forest reserves, relaxed his opposition and agreed not only to support the transfer, but to introduce and sponsor such a bill.[48] Pinchot's friendship with Westerners was beginning to pay off. Mondell's first attempt, in the winter of 1903–04, failed because of opposition from the Homestake Mining Company, which feared that it would lose timber privileges in the Black Hills, South Dakota, Forest Reserve.[49] Modified to satisfy the

[46] The first recommendation for transfer in the annual reports of the General Land Office was in 1904, the first report written by Richards. See *Annual Report of the General Land Office, 1904*, 50–51.

[47] Pinchot to Newell, Aug. 30, 1902 (GP #80); Pinchot to Maxwell, Aug. 30. 1902 (GP #80); Pinchot to Robert U. Johnson, June 14, 1903 (GP #79).

[48] Representative Mondell's change in attitude is one curious feature of the transfer legislation. Early in 1903 he had fought the transfer bill in Congress, but by October 5, 1903 he had changed his mind and was willing to support it. See Pinchot to Henry S. Graves, Oct. 5, 1903 (GP #79). Almost a month earlier Roosevelt indicated that Western leaders had changed their views; see Roosevelt to Pinchot, Sept. 11, 1903 (TR). Although the precise explanation for this change is not clear, it appears to be related to specific land policy in Wyoming and the fact that Pinchot favored a land use program more in accord with the desires of Wyoming groups than did the Department of the Interior.

[49] Senator Kittridge of South Dakota, who was "practically the whole opposition to the transfer," represented the interests of the Homestake Mining Company. See Pinchot to Roosevelt, July 15, 1904 (GP #84). Pinchot tried to bring pressure to bear on Kittridge in various ways. See Pinchot to Seth Bullock, Mar. 15, 1904 and Mar. 17, 1904; Bullock to Pinchot, Mar. 16, 1904 (all in GP #82). The mining company feared that it would lose its preferred position in purchasing timber from the Black Hills reserve. It was satisfied

Company, the bill passed readily at the following session of Congress and became law on February 1, 1905.

Now fully in charge of the reserves, Pinchot immediately set out to perfect and expand the national forest program. While the General Land Office had construed its powers narrowly, Pinchot hoped to accomplish more by a broader interpretation of existing legislation. According to the Forest Management Act of 1897, the Secretary of the Interior could "make such rules and regulations and establish such services as will insure the objects of [the] reservations, namely, to regulate their occupancy and use and to preserve the forests therein from destruction." This Act, Pinchot argued, conferred upon the Secretary "every necessary authority and power for their [the reserves'] management by whatever methods he may deem best. Legally there is no obstacle to the introduction of the most practical and approved ways of handling forest lands."[50] This broad interpretation of the Act of 1897 played a crucial role in the new forest program by enabling the administration to justify many measures which Congress might not have sanctioned. As Pinchot's chief law officer later remarked, "The vitalizing of this power through vigorous use was the chief means whereby the Forest Service achieved results in matters of grazing, water power, the prevention of land frauds, etc. Comparatively little conservation legislation was enacted during these years. Progress came not through getting new powers . . . but by using those we had."[51]

Law officers of the Forest Service played a crucial role in the new forest program. As early as 1903 Pinchot brought into the Bureau of Forestry, as chief legal adviser, his former Yale classmate, George Woodruff, who became the key figure in the development of the legal and constitutional rationale for the Roosevelt conservation movement.[52] A native of Pennsylvania, Woodruff graduated from

with a provision in the transfer act that green timber could not be exported from South Dakota. See memorandum written by Paul Kelleter, Nov. 27, 1912 (GP #60).

[50] *Department of Agriculture, Yearbook, 1899*, 297–298.

[51] Philip P. Wells, "Personal History," MSS in GP #1671.

[52] Pinchot to Woodruff, Feb. 2, 1903 (GP #81).

Yale in 1889, and received his L.L.B. from the University of Pennsylvania in 1895. He served with Pinchot in the Forest Service until 1906, where he developed the legal basis for the new policies of requiring fees for grazing and permits for hydroelectric power installations in the national forests. From 1907 until 1909 he served as Assistant Attorney-General, assigned to the Department of the Interior. In later years he followed the Chief Forester to Pennsylvania, where he became Attorney-General of the state during Pinchot's governorship from 1923–27, and a member of the Public Service Commission of Pennsylvania from 1931 until his death in 1934. A fellow lawyer in the Forest Service later recounted Woodruff's crucial contribution to the Roosevelt conservation program: "The Act of June 4, 1897 had placed us almost in the legal position of the agent of a private landowner with very broad powers to manage the property for the owner's welfare. The principles of real property law were available to use for the protection of the public interests. Woodruff at once seized upon them and applied them to the new conditions."[53]

An expanded forest program required increased revenue. The new Forest Service received not only the customary appropriation for national forest protection, but, in addition, the funds for the Bureau of Forestry. Pinchot hoped to increase these sums even further by charging rentals for grazing and water power installations, and by increasing the price of standing timber.[54] The Forest Service, for example, soon replaced competitive timber bidding with sale at a figure closer to the appraised price. Between 1900 and 1904 the Department of the Interior, selling to the highest bidder, received an average price of $1.15 per thousand board feet.[55] In supervising sales on the Chippewa Indian lands in Minnesota, however, the Forest Service had received an average of $7.00 per thousand.[56] Similar results, Pinchot argued, could be achieved with the national

[53] Wells, "Personal History," (GP #1671).

[54] Roosevelt to Thomas M. Patterson, Dec. 21, 1905 (TR).

[55] This figure is compiled from the Annual Reports of the Commissioner of the General Land Office from 1897 to 1905.

[56] *Forestry and Irrigation* (Feb. 1904), 10, 53–54.

forests. Soon after the transfer, Eugene Bruce, an Adirondack lumberman who came to the Forest Service in 1902, became chief trouble-shooter on timber sales contracts. His work rapidly increased returns from timber sales. For example, in the spring of 1905, on a contract for diseased white pine in the Black Hills Reserve, he brought about a price increase from 25 cents per thousand to 75 cents. Shortly thereafter he raised the price in a timber contract in the Big Horn National Forest from $1.00 to $2.50 per thousand, the first large federal timber sale over $1.00. In one case Bruce's negotiations broke an agreement between two competitors for a maximum of $2.00 and produced a final price of $5.05.[57]

The Forest Service wished to obtain greater control over its revenue as well as to increase its amount. Prior to 1905, forest receipts were deposited in the General Treasury, subject to congressional appropriation for any purpose. Pinchot hoped to reserve this revenue specifically for the Forest Service. In the Transfer Act of 1905 he secured a provision creating a special forestry fund. For five years all proceeds from the sale of national forest products would be deposited in this fund and made available for the Secretary of Agriculture to spend at his discretion. Pinchot patterned the Forestry Fund after the Reclamation Fund. In 1907 he attempted to make this arrangement permanent in a bill which provided that it remain until Congress chose to discontinue it. His effort backfired; the lawmakers responded by abolishing the fund entirely.[58]

Pinchot reorganized the Forest Service and infused it with a new spirit of public responsibility. He decentralized its administration and gave more authority to its field officers. With his indomitable

[57] This change is fully covered in E. S. Bruce, "Recollections of Pioneer Forestry," MSS in GP #1714. On Dec. 1, 1908 the Department of the Interior began to charge market value rather than $2.00 per acre for forest lands under its jurisdiction.

[58] Peffer, *Public Domain*, 91–96, describes the forestry fund problem. One reason for Pinchot's desire to ensure the forestry fund was a fear that Western congressmen would divert proceeds from the sale of forest products to other Western enterprises. For example, in 1906 Newlands planned legislation to apply the funds derived from the sale of timber to the reclamation fund. See Newlands' secretary to Charles Walcott, Mar. 19, 1906 (FGN, Letters).

energy and enthusiasm he injected a missionary spirit into the young foresters who came into the Service, a spirit witnessed by their intense devotion and respect for their "Chief" over the years.[59] Through such innovations as these, the Forest Service developed from a coterie of law clerks into a well-trained force fighting for forest protection and more scientific forest management.

The Roosevelt administration greatly enlarged the area of the national forests. When Roosevelt became president 41 reserves had been set aside totaling 46,410,209 acres; in his first year as president he created 13 new forests of 15,500,000 acres. From this time until his re-election in 1904, Roosevelt created few new reserves, but in 1905 he again took up the task. In 1907, in reaction to the President's exercise of his executive discretion in resource policy, Congress revoked his authority to create reserves in six Western states. The President, however, spoke the last word. Between the time that Congress passed the measure and he signed it, Roosevelt set aside 75,000,000 additional acres in reserves, increasing the total to 150,832,665 acres in 159 national forests.[60]

Forest enthusiasts soon called upon Congress to extend the national forest system to the entire country by adopting a new policy of purchasing Eastern land for reserves. As early as 1885, park organizations of the southern Appalachians proposed a national forest in that area, and fifteen years later New Englanders created the Society for the Preservation of New Hampshire Forests to spearhead a drive for a similar national reservation in the Northeast.[61] For several years these two groups carried on their campaigns separately, but in 1906 they joined forces to push their common cause. Although Roosevelt and Pinchot at first discouraged these proposals, by 1906 they gave the movement their influential support

[59] See praise for Pinchot by one of his later opponents in Greeley, *Forests and Men*, 64–86.

[60] These figures are compiled from the annual reports of the Secretary of the Interior and the Secretary of Agriculture.

[61] Ise, *U. S. Forest Policy*, 207–223 covers this entire movement. For the early proposal for the park see *The Forester* (Dec. 1899), 5, 283; (July 1900), 6, 160.

and the law officers of the Forest Service prepared a measure to create the Eastern forest.[62] Relying on the constitutional power of the federal government to regulate navigable streams, the bill provided for the purchase of timberlands which if protected would improve the navigability of rivers with headwaters there. Controversy over the constitutionality of the measure held it up, but in 1908 a Senate Judiciary Committee gave its assent. On March 1, 1911 Congress passed the Weeks Act, under which the federal government acquired large areas of Eastern land for national forests and thereby applied the Forest Service program to the entire country.

[62] Prior to this time Roosevelt had discouraged efforts to create an Appalachian National Forest, and had urged that the states purchase the lands themselves. See Roosevelt to William S. Harvey, Sept. 16, 1905 (TR). Roosevelt's attitude may have depended on Pinchot's views. The latter had been opposed to the White Mountain Reserve in New England, but came to realize that New England support was essential if the southern Appalachian forest, which Pinchot favored, was to win approval. See penciled note by Philip W. Ayres, dated 1905, in Society for the Protection of New Hampshire Forests, *Scrapbook, 1901–1910*. See also Wells, "Personal History" (GP #1671).

Chapter IV

Range Wars and Range Conservation

The federal irrigation and forestry programs inspired the Roosevelt administration to take up grazing problems, a task which revealed even more closely than did forestry the close collaboration between conservationists and large-scale businessmen. Many in the West demanded that the national forests include large areas of range land in order to increase their value for watershed protection. Grazing, moreover, became the primary commercial use of the forests, far more important than lumbering.[1] The Forest Service, looking upon grazing with greater favor than did the Secretary of the Interior, opened the reserves to more livestock and took a keen interest in the entire Western grazing industry. This new emphasis prompted the administration to formulate policies to cover public range outside as well as within the forests, and thereby, as allies of the large cattle corporations, to become embroiled in the perennial bitter conflicts of the cattle country.[2]

Chaos and Control on the Range

Much of the Western livestock industry depended for its forage upon the "open" range, owned by the federal government, but free

[1] Only in the Black Hills Reserve, in South Dakota, where the Homestake Mining Company used considerable timber, was lumbering most important. Pinchot constantly emphasized the great importance of grazing. See Pinchot, *Breaking New Ground*, 268; *Proceedings, American National Livestock Association, 1901*, 274; *Forestry and Irrigation* (Nov. 1901), 7, 276; Peffer, *Public Domain*, 74.

[2] Peffer, *Public Domain*, 3–168, is the best treatment of the range question.

for anyone to use.[3] Moving their livestock from the higher alpine ranges during the summer to the lower grazing lands in the winter, cattle and sheepmen could operate profitably with little capital and no privately owned land. Chaos and anarchy, however, predominated on the open range. Congress had never provided legislation regulating grazing or permitting stockmen to acquire range lands. Cattle and sheepmen roamed the public domain, grabbing choice grazing areas before others could reach them first. Cattlemen fenced range for their exclusive use, but competitors cut the wire. Resorting to force and violence, sheepherders and cowboys "solved" their disputes over grazing lands by slaughtering rival livestock and murdering rival stockmen. Armed bands raided competing herds and flocks and patrolled choice areas to oust interlopers. Absence of the most elementary institutions of property law created confusion, bitterness, and destruction.

Amid this turmoil the public range rapidly deteriorated.[4] Originally plentiful and lush, the forage supply was subjected to intense pressure by increasing use. The number of Western cattle grew rapidly after the Civil War; a rising sheep industry claimed its right to share in the public range; and settlers transformed grazing lands into more valuable cropland. The public domain became stocked with more animals than the range could support. Since each stockman feared that others would beat him to the available forage, he grazed early in the year and did not permit the young grass to

[3] Ernest Staples Osgood, *The Day of the Cattleman* (Minneapolis, 1929), 176–258, contains the best account of the effect of the absence of range legal institutions.

[4] Will C. Barnes, *Story of the Range* (Washington, 1925), 6–9; Edward H. Graham, *Natural Principles of Land Use* (New York, 1944), 143–160. The amount of range deterioration is a highly controversial question. Stockmen frequently maintain that range quality has not declined, but in the Progressive Era they often expressed fear that continued intensive use would destroy the entire industry. See *Proceedings, American National Livestock Association, 1898*, 94–100; *1899*, 188–191. In 1944 the U. S. Forest Service estimated that public domain range lands were depleted 67 per cent from their virgin condition, privately owned lands 51 per cent and national forest lands only 30 per cent; 74th Congress, 2nd Session, *Sen. Doc. 199, The Western Range*, 108–116.

mature and reseed. Under such conditions the quality and quantity of available forage rapidly decreased; vigorous perennials gave way to annuals and annuals to weeds.

An interest in more stable range conditions coincided with the collapse of the booming, speculative, and unstable livestock industry which had grown up in the West in the 1870's and 1880's. A rapidly rising demand for meat in the urban centers of Europe and the United States, and the perfection of refrigerated shipping attracted many capitalists to the Western range in the late 1870's. During the peak years from 1879 to 1885 these corporations reaped huge profits.[5] But the market soon became saturated, forage grew scarce, and many stockmen went bankrupt. The crisis came during the severe winter of 1886–87 when hundreds of thousands of livestock died from cold and starvation. The decline had begun prior to that time, but the catastrophe of the blizzard winter put the finishing touch to a collapsing, competitive industry, and marked the beginning of a new departure which emphasized permanent, long-term management and continuous, though smaller, profits. To ensure more dependable feed supplies, stockmen began to combine summer range with winter hay raised on irrigated farms in the river valleys. Through selective breeding and disease eradication they improved the quality of their animals. They transformed a migratory industry into one requiring larger investment and operating from a permanent, home base onto a more restricted range.[6]

Federal laws did not encourage these efforts to stabilize the livestock industry. Under the homestead acts a settler could acquire sufficient land to raise hay and establish a farmstead, but he could not secure control of federal grazing land for summer use. Many solved this problem by leasing range owned by railroads or the states.[7] Those who depended on federal lands, however, could not

[5] Herbert O. Brayer, "The Influence of British Capital on the Western Range-Cattle Industry," *Journal of Economic Industry*, Supplement (1949), IX, 85–98.

[6] Osgood, *Day of Cattleman*, 216–237.

[7] In 1900, according to one official, the demand for state and railroad leases exceeded the supply by five million acres. See *Proceedings, American National Livestock Association, 1899*, 175–177; *1900*, 266.

exclude competitors who might obtain the lion's share of the grass. Cattlemen had tried to solve this problem by fencing portions of the federal range to exclude both sheepmen and settlers, but in 1885 Congress declared this practice illegal.[8] In response, cattlemen proposed that Congress permit homestead entries with sufficient land to support stock herds, or cede the public domain to the states which then would add it to their lands already under lease.[9] These pleas fell on deaf ears. Yet the cattlemen did not give up, and from the 1880's on they pushed a measure providing that the federal government retain ownership and lease the range to the stockmen.[10]

The federal leasing measure also appealed to scientists in Washington who were concerned with the progressive deterioration of the range, and sympathized with attempts to develop a more stable and permanent industry.[11] In 1895 the Secretary of Agriculture organized in his Department a Division of Agristology, the first federal agency to consider range management. In 1901 this Division merged with the Division of Botany to form the Bureau of Plant Industry, and two years later the Bureau established the Santa Rita Range Reserve in southern Arizona to demonstrate the value of range protection in increasing forage yields. Close cooperation between federal agencies and state experiment stations produced studies in the growth, care, reseeding, and managment of pastures.

[8] Osgood, *Day of Cattleman*, 187–195. Many writers, e.g., Roy Robbins, *Our Landed Heritage* (Princeton, 1942), 285–298, interpret these violations as merely fraudulent actions by large corporations. It is far more significant that fencing was one among many alternatives devised by stockmen to facilitate a more stabilized industry in the face of inadequate land laws.

[9] *Proceedings, American National Livestock Association, 1899*, 83–94; *1900*, 83, 141–145.

[10] See, for example, *Ibid., 1903*, 247–249.

[11] For a brief summary of this development see L. A. Stoddart, "Range Management," in *Fifty Years of Forestry in the U. S. A.* (Washington, 1950), 113–135. See also Jared G. Smith, *Grazing Problems in the Southwest, U. S. Department of Agriculture, Division of Agristology, Bulletin #16*; Arthur W. Sampson, *The Revegetation of Overgrazed Range Areas, Preliminary Report, United States Department of Agriculture, Forest Service Circular #158*.

From this experimental work state and federal range experts evolved a forage improvement program which emphasized the correct timing of range-use to facilitate natural grass seeding, and the adjustment of animal populations to the carrying capacity of the range. Although they hoped to apply these principles to both private and public lands, federal scientists were most concerned with the serious condition of the public domain. Since their program depended on strict control of the number of animals using the range, they enthusiastically backed the stockmen's leasing plan.

Proposals to stabilize the use of the public range became closely involved with Western economic and political conflicts. For many years cattle and sheep operators on the one hand, and both groups and farmers on the other had struggled for control of the public grazing lands. New settlers, arguing that grazing should give way to farming, staked out claims in the cattle country. Cattlemen replied by fencing their lands or by taking up all available water rights, without which settlers could not farm. Often armed conflicts arose between the two groups. In the Johnson County, Wyoming "war" of 1892, for example, cattlemen invaded a stronghold of settler resistance, murdered two settlers, found themselves besieged on a ranch and were saved in the nick of time by a detachment of federal troops.[12]

These conflicts frequently erupted in state and federal politics. State and community boosters who hoped to encourage a growing population sided with the settlers; the livestock industry, they argued, should give way to a more populous farming community. They established state immigration commissions to attract people from the East and lobbied in Washington for laws which would enable newcomers to acquire Western land more easily. Fearing that any measure to aid the cattlemen would withdraw lands from entry and retard settlement, these groups at every turn opposed fencing and leasing. In self-defense the cattle owners tried to control state politics and to thwart federal homestead legislation. They incurred the hatred of those promoting settlement to such an extent

[12] Osgood, *Day of Cattleman*, 204–205, 213, 237–255.

that Western politicians could espouse the cause of the cattle "barons" only at the risk of their political lives.[13]

In the 1880's cattlemen faced a second competitor as sheep from the Northwest and Southwest moved into the mountain states. Far more mobile than cattle, sheep provided a formidable threat to the cattlemen's range. Almost at a moment's notice a sheepherder could move his flock to a new site and take advantage of choice forage more quickly than could cattle which roamed without guidance. Moreover, according to the cattlemen, their stock would not graze where sheep had grazed and the smell of sheep remained.[14] Since sheep were continually on the move, ranging from one state to another, the more stationary ranchers, and, in fact, most settled Western communities, looked upon them as intruders. The prevalence of Basques among the herders gave the industry an even more alien character.[15] Cattlemen obtained state laws prohibiting sheep grazing near villages and towns and heavily taxing out-of-state flocks. They hoped that the leasing proposal would exclude sheep from their ranges by giving preference to local, established ranchers. For this very reason the sheepmen bitterly opposed efforts to include grazing land in the national forests or to adopt a leasing program for the public domain.[16]

These conflicts over control of the public range confronted the Roosevelt administration with a serious dilemma. To carry out its plans for a sustained-yield range program through federal leasing it would be forced to rely on the large cattle corporations for political support. This, however, would alienate the small farmer and deprive the administration of its most important source of Western backing. The necessity for range conservation tipped the scales in favor of the cattlemen. To adopt the views of the smaller farmers would

[13] See, for example, Anne Carolyn Hansen, "The Congressional Career of Senator Francis E. Warren from 1890 to 1902," *Annals of Wyoming* (Jan. 1948), 20, 5–16.

[14] J. W. Dorsey to Francis G. Newlands, Dec. 20, 1901 (FGN, Letters).

[15] J. W. Freeman to Newlands, nd (between January and May 1902) (FGN, Letters).

[16] *Proceedings, American National Livestock Association, 1899*, 141–150, 175–198; *1900*, 184–190, 260–312.

have prevented the administration's technicians from even approaching their goal of better range management. The decision to support the leasing measure proved to be a fateful one. It identified administration leaders with the large cattle corporations, so far as the West was concerned, and brought down upon their heads the wrath of farming groups and their representatives in Congress. It marked the first sustained attack on the administration's conservation program, created a rupture in the coalition which between 1903 and 1905 Pinchot had formed to support his forest policies, and played a crucial role in the refusal of Congress in 1908 and 1909 to go along with Roosevelt's resource plans.

Protecting the Forest Ranges

In the 1890's the Department of the Interior, not recognizing grazing as a legitimate use of the forest reserves, sought to exclude cattle and sheep from them entirely. In April 1894 the Secretary of the Interior issued an order forbidding driving, feeding, grazing, pasturing, or herding cattle, sheep, or other livestock in the forests.[17] The Secretary had no power to enforce this measure, and the Army, in charge of patrolling the reserves, refused to execute it on the grounds that it was illegal. But vigorous protest from stockmen moved the Department to undertake a closer investigation of grazing conditions to determine the reasonableness of its order. In 1897 Frederick V. Coville, botanist of the Department of Agriculture, carried out in the Western reserves the first scientific range investigation in the United States. Coville urged that the Department adopt a more varied policy. In public recreation areas and on the major reservoir watersheds, he argued, grazing should be prohibited, but in many other areas the Department should permit it, though under strict regulation.

Frederick V. Coville deserves special notice. Although he never achieved the public prominence of his fellow conservation scientists, Coville played an extremely important role in formulating the

[17] The Secretary issued his grazing order after receiving petitions from irrigators, especially in western Colorado. *Proceedings, American National Livestock Association, 1899,* 205.

scientific basis for a range conservation policy. Technician rather than promoter, he remained in the background while Newell, Pinchot, and others moved into the spotlight. Coville spent almost his entire professional career in the federal service. After graduating from Cornell University in 1887 he remained there for a year as a botany instructor. In 1888 he went to Washington to become Assistant Botanist for the United States Department of Agriculture, and in 1893 was promoted to Chief Botanist. For many decades he was a leader in Washington scientific circles, serving as a member of the board of managers of the National Geographical Society and president of the Biological Society of Washington and the Washington Academy of Sciences. From the beginning of his public botanical work, Coville took a keen interest in Western grazing problems. Pinchot frequently called upon him for technical advice in dealing with the forest ranges, and when the Public Lands Commission took up the task of formulating a range policy for the entire public domain in 1903, it also relied heavily on Coville's investigations and recommendations.[18]

Following Coville's suggestions, in 1898 the Department of the Interior issued cattle grazing permits for most of the forest ranges, but confined sheep permits to Oregon and Washington.[19] The Department agreed with irrigators that sheep especially damaged the range; they cropped the grass very close to the ground, tramped it with their hooves, killed it where they bedded down for the night, and ate tree seedlings. Sheepherders, the federal administrators argued, were responsible for most forest fires. The Department, however, finally yielded even to the sheepmen. After receiving numerous petitions from stockmen and detailed reports on grazing conditions from forest superintendents, in 1899 it opened ten reserves to sheep grazing.[20] The Department approved these changes only with great reluctance. Ethan A. Hitchcock, who became Secre-

[18] A brief sketch of Coville's work is in Stoddart, "Range Management," 118.

[19] U. S. Department of the Interior, *Annual Report of the Commissioner of the General Land Office, 1898* (hereafter cited as *Ann. Rept. GLO*), 87.

[20] *Ann. Rept. GLO, 1899*, 107–108, 110.

tary of the Interior in 1899, would have preferred to prohibit grazing.[21] The Commissioner of the General Land Office warned, in his code of grazing regulations, that stockmen used the forests only as a privilege and not as a right, and that the Secretary could exclude them entirely at his discretion.[22]

Gifford Pinchot sharply challenged these views; it was he who led the fight in Washington for greater recognition for the stockmen. In the Bureau of Forestry, Pinchot established a Division of Grazing, and in 1901 he brought an Arizona sheepman, Albert F. Potter, to Washington as its head. Potter's role in federal range affairs symbolized the new working relation between Western stockmen and the Roosevelt administration. A Californian by birth, Potter went to Arizona in 1882 for his health, and entered the livestock business. At first a cattleman, in 1895 he shifted to sheep raising and became one of the most prominent sheepmen in the territory. Active in organizing the Arizona Wool Growers' Association, he served as its secretary from 1898 to 1900. In this capacity he came into close contact with Pinchot over the issue of permitting sheep grazing in the Arizona forests. The two men saw eye to eye on that question, and Pinchot well understood the political wisdom of cementing his relationship with the entire Western grazing industry through Potter. Whereas Coville was the technician, and Pinchot the promoter, Potter was the politician of the new range policy. He became one of the most influential figures in the federal forest administration. He took charge of grazing when the national forests were transferred to the Department of Agriculture. In the short period between Pinchot's dismissal as Chief Forester in January 1910, and the appointment of his successor, Henry Graves, Potter served as acting Chief. From that time on he was Graves' assistant,

[21] Peffer, *Public Domain*, 72–73.

[22] The General Land Office *Manual* read, "The Secretary of the Interior, in being charged with the proper protection of the forest reserves, has the right to forbid any and all kinds of grazing therein." In contrast, the Forest Service regulations stated, "The Secretary of Agriculture has authority to permit, regulate, or prohibit grazing in the forest reserves." See Pinchot, *Breaking New Ground*, 265.

as Associate Chief Forester, and continued to speak for the stockmen in the Forest Service.

Opening the forest reserves to grazing created a series of controversies between the stockmen and the federal government which were not easily resolved, and which illustrated the political implications of the conservation program. For example, the old conflicts between farmers and cattlemen and between cattle and sheepmen remained, but their focal point now shifted to forest officials who became deeply entangled in intra-Western strife. Who should receive priority to use the range? Each stockman now had to convince a departmental official that his case was just; conflicts between rival users were settled by administrative decisions. Those who obtained favorable rulings from the Forest Service supported its work, but those who did not complained bitterly of "bureaucratic tyranny" and challenged the constitutionality of the grazing rules. In a test case, decided May 3, 1911, the Supreme Court upheld the validity of the Forest Management Act of 1897 and the power of the Department of Agriculture to administer the forest range.[23] But this hardly spared the Forest Service from criticism from the stockmen. That agency, continuing to serve as an arbiter among competing resource users, could not please everyone.

When the Department of the Interior inaugurated the grazing permit, it prohibited out-of-state herds and flocks from entering any reserve. This, however, interfered with established patterns of migratory grazing across state lines. Livestock which used winter range on the Red Desert in southwestern Wyoming, for example, could no longer migrate to summer pastures in the Uinta Mountains in northeastern Utah. Wyoming stockmen, therefore, persuaded Senator Clark of that state to introduce a bill to abolish the rule. Although Clark failed, the Secretary of the Interior modified the regulations and in January 1902 instituted a more elaborate system of priorities: residents in the reserve, those owning permanent stock ranges in the reserve but residing outside, those living in the

[23] *United States* v. *Grimaud,* 220 U. S. 506.

vicinity and owning "neighboring stock," and outsiders who could present some equitable claim.[24]

This policy favored the stock operator whose home base lay near his pasture lands. Stockmen from far away, especially in the sheep industry, more migratory than cattle raising, received lower priorities. In fact, resident cattlemen, looking upon the reserves as a means to protect themselves from competition from sheep flocks, often petitioned the Department to establish new reserves or extend old ones, and thereby to increase the area of range which they could use exclusively.[25] Sheepmen, in turn, protested against these changes.[26] As federal officials established priorities, however, which continued from year to year, the conflicts subsided. Since the relatively secure use of regulated forest range contrasted greatly with the insecurity of grazing on the unregulated free range of the public domain, those who used the forests enthusiastically supported the federal program.

The decision to permit grazing in the forests raised an even thornier problem: should the stockmen pay a fee for using the range? As early as 1899 several forest supervisors had suggested this innovation. The General Land Office did not believe that existing legislation permitted a fee, but recommended that Congress authorize it.[27] Representative John F. Lacey of Iowa twice introduced such a measure in the House, but it never came out of committee. A variety of motives prompted those who argued for a grazing fee. Some believed simply that the federal government had no business giving away its resources. The Department of the Interior hoped that the added revenue would prompt Congress to appropriate more funds for its work. The fee might also solve a vexing administrative

[24] *Ann. Rept. GLO, 1902*, 100.

[25] Colin B. Goodykoontz, *Papers of Edward P. Costigan Relating to the Progressive Movement in Colorado 1902–1917* (Boulder, 1941), 244; Roosevelt to Pinchot, Apr. 9, 1906 (TR). Cattlemen also protested the elimination of lands from the reserves.

[26] "Proceedings of Conference Between Special Land Commission Appointed by President Roosevelt and Prominent Stockmen of the West," in *Proceedings, American National Livestock Association, 1905*, 301–304.

[27] *Ann. Rept. GLO, 1899*, 110.

problem. When the Department had brought action against trespassers on the forests in violation of grazing regulations, lower courts had ruled against it. No law permitted appeal in such cases.[28] Injunctions to prevent trespass were slow and cumbersome but the Department could secure effective results if each stockman, in paying a fee, was forced to take out a permit.[29]

As far as Pinchot was concerned the Forest Management Act of 1897 had already granted the administration the necessary power to establish grazing fees. Although that Act did not specifically authorize a grazing charge, it did not prohibit it, and, argued Pinchot, the broad grant of authority "to regulate [the] occupancy and use" of the forests included permission to establish a fee.[30] After Congress transferred the reserves to the Department of Agriculture, Pinchot obtained from the Attorney-General an interpretation of the law which upheld his view. In January 1906 he imposed the first grazing fee,[31] and five years later the Supreme Court approved the innovation as a proper exercise of administrative authority.[32]

The Public Domain Range

The Forest Service could perfect its grazing policy by expanding the interpretation of laws which Congress had already passed. To apply this program to the public domain, however, required new legislation establishing federal control outside the forests. The administration found a ready solution to the problem in the leasing measure which the cattlemen had long since proposed. After 1906, Roosevelt, Pinchot, and James R. Garfield, who became Secretary

[28] The legal aspects of grazing control are summarized in Ise, *U. S. Forest Policy*, 168–173.

[29] Philip P. Wells, MSS, "Personal History" (GP #1671).

[30] U. S. Department of Agriculture, *Yearbook, 1899*, 297–298.

[31] Pinchot carefully prearranged Attorney-General Moody's decision of May 31, 1905. Learning that Moody might render an adverse decision, Pinchot discussed the matter with him and convinced him that the fee was valid. Pinchot, *Breaking New Ground*, 272.

[32] *United States* v. *Grimaud*, 220 U. S. 506, decided May 3, 1911. See Ise, *U. S. Forest Policy*, 168–173. An account of this problem by the forest supervisor who arrested Grimaud is in Greeley, *Forests and Men*, 79–80.

of the Interior in 1907, rallied behind this measure as the best method of promoting range conservation.

Representative Bowersock of Kansas introduced the first major leasing bill in Congress in 1901.[33] It encountered immediate opposition. According to Secretary of the Interior Hitchcock, hostile to the cattlemen, it would encourage land monopoly and defeat the purposes of both the homestead laws and the proposed reclamation measure.[34] Members of the House Public Lands Committee supported Hitchcock; by tying up the public lands, they argued, grazing leasing would retard settlement.[35] Although the Bowersock measure failed, Congress passed a compromise, the Kincaid Act of 1904, which authorized 640-acre homestead patents in western Nebraska.[36] To many in Congress the Kincaid Act offered a possible solution; if successful in Nebraska it could be extended to the entire public domain. The origin of the Kincaid Act illustrated the controversy involved in federal leasing. Cattlemen had proposed to the Nebraska legislature a memorial to Congress backing a full-scale leasing program for the public domain. In Nebraska, however, representatives of the farmers had vigorously opposed this move, and had engineered a compromise memorial to Congress asking for a 640-acre homestead act. The cattlemen looked upon this as a defeat, protested against the memorial, and complained that such a law could not possibly solve their problem. The Kincaid Act, therefore, represented a victory for the settlers over the cattlemen.[37]

Agitation for leasing, coupled with a growing concern over operation of other land laws, prompted the administration to appoint a Public Lands Commission in the fall of 1903. One of its primary

[33] Col. John P. Irish drew up this bill for the American Cattle Growers Association, an organization of California and Nevada cattlemen. See Charles Greene to Newlands, Feb. 4, 1902, with enclosed news clipping (FGN, Letters).

[34] U. S. Department of the Interior, *Annual Report of the Secretary, 1902*, 167–174.

[35] Newlands to A. W. Riley, Jan. 23, 1902 (FGN, Letters).

[36] Arthur R. Reynolds, "The Kincaid Act and Its Effect on Western Nebraska," *Agricultural History* (Jan. 1949), 23, 20–29.

[37] *Proceedings, American National Livestock Association, 1905*, 291–292.

objectives, so far as Pinchot was concerned, was to investigate grazing conditions on the public domain, and to formulate proposals to stabilize the range. In the West this Commission once more stirred up a hornets' nest. In January 1904 it met with stockgrowers in Portland, Oregon to discuss the range, and later held sessions in other Western centers, culminating in a large gathering in August in Denver.[38] At these meetings the large cattle corporations spoke for leasing and representatives of the smaller groups spoke against it. Following the proceedings closely, Western newspapers frequently condemned the leasing proposition as monopolistic and a threat to further homesteading.[39] The Commission studied the practical effects of overgrazing. In order to determine the feasibility of establishing leasing on the public lands, it requested Frederick V. Coville to examine the leasing systems used by the Northern Pacific Railroad, the state of Texas and the state of Wyoming.[40] In its final report the Commission advocated classification of the grazing lands and their management as grazing districts.[41]

Prior to the appointment of the Public Lands Commission, the Roosevelt administration had not yet presented to the country a definite public domain grazing policy. In the spring of 1903 Pinchot leaned toward the leasing proposal, but he hesitated to speak out openly, and frequently expressed the farmers' point of view.[42] After his more intimate contact with the range as a member of the Public Lands Commission, he supported leasing with more enthusiasm. By speeding up prosecutions for illegally erecting fences on the public domain, the administration itself prodded cattlemen

[38] For a full report on this meeting see *Ibid.*, 277–364.

[39] See, for example, *Great Falls* (Montana) *Tribune*, Aug. 23, 1904 (FGN, Scrapbook #18).

[40] Public Lands Commission, *Report*, 27–61.

[41] *Ibid.*, xxii–xxiii.

[42] Pinchot to R. W. Gilder, Mar. 10, 1903 (GP #79). This reluctance depended less on conviction as to range conditions, and more on Pinchot's desire to accomplish the transfer before he supported a policy unpopular to many in the West. In 1901 he wrote Representative Malcolm A. Moody of Oregon that leasing was needed, but that it was not yet time to make it a public issue. Peffer, *Public Domain*, 74.

to push the leasing measure more forcefully.[43] Both Pinchot and Roosevelt fully realized that to support leasing would create Western opposition and identify the administration with the large cattlemen. "The people," complained the President, "refuse to face squarely the proposition that much of these lands ought to be leased and fenced as pastures, and that they can not possibly be taken up with profit as small homesteads."[44] The administration carefully waited until after the election of 1904 and the successful transfer of the reserves to the Bureau of Forestry on February 1, 1905 before undertaking such a hazardous course of action. Then, in December 1905, in his message to Congress, the President squarely backed the measure. Methods practiced on the forest reserves, he argued, Congress should adopt for the public domain.

In formulating its proposal the administration conferred frequently with leading Western cattlemen, and called a conference in Washington to obtain their advice.[45] It then attached a leasing plan to the 1907 appropriations bill and, for the first time, brought it to the floor of the House for full-scale debate. Western congressmen in Washington bitterly denounced the bill, while public opinion in the West itself burst forth in criticism.[46] Late in 1906 the Colorado legislature urged its governor to call a special public lands convention to make a unified protest to the administration.[47] Although Congress had defeated the measure by the time the convention assembled in June 1907, feeling among sheepmen and farmers still ran high.[48] The meeting produced no significant immediate results,

[43] Roosevelt to J. C. Underwood, Dec. 26, 1906 (TR); Roosevelt to Francis E. Warren, Feb. 11, 1907 (TR).

[44] Roosevelt to Wm. B. Lighton, Apr. 6, 1906 (TR).

[45] Roosevelt to F. M. Stewart, Jan. 30, 1907 (TR); Roosevelt to E. J. Bell, Feb. 4, 1907 (TR).

[46] Peffer, *Public Domain*, 90–91.

[47] *Forestry and Irrigation* (June 1907), 13, 278. Many writers, especially Ise, *U. S. Forest Policy*, 174–175, interpret this convention as a reaction to Roosevelt's arbitrary creation of forest reserves in the face of congressional prohibition against them. The facts that the convention was proposed in December 1906, and the reserves were created in March 1907 disprove this view.

[48] Peffer, *Public Domain*, 99–102.

but it revealed the dilemma into which the administration had plunged when it had decided to take up leasing on the public domain. Fully aroused, President Roosevelt sent to the convention a contingent of cabinet members. On the convention floor prominent Republicans, among them Senators Francis Warren of Wyoming and Reed Smoot of Utah, defended the Roosevelt program. Representatives of the American National Livestock Association, who upheld the Forest Service grazing rules and the public domain leasing measure, gave added and full support. A few weeks after the meeting, Roosevelt wrote to the president of that Association, "I deeply appreciate . . . all you did at the convention."[49]

The grazing venture threatened to alienate from the administration the settler and homestead groups which had rallied behind the federal irrigation program. Stockmen and irrigators had often come to blows in the West. Cattlemen had vigorously opposed almost every phase of the irrigation movement in the late 1880's and early 1890's because it would divert rangeland into cropland.[50] Irrigators, on the other hand, had been none too happy when the Secretary of the Interior opened the forests to grazing, and they fought the measure to lease the public domain. The leasing measure severed relations temporarily between George H. Maxwell and the administration. In 1904 Maxwell, for example, refused to open his section of the National Irrigation Congress to a paper on range management to be read by Prof. R. H. Forbes, Director of the Arizona Agricultural Experiment Station. Forbes appeared on the program, but under the forestry section, of which Pinchot himself was in charge.[51] Not until the fight over multiple-purpose river development which began in 1908 did Maxwell again identify himself closely with the inner group of conservation promoters.

The new Secretary of the Interior, James R. Garfield, in contrast

[49] Roosevelt to Governor Henry A. Buchtel, May 23, 1907 (TR); Garfield to Pinchot, May 27, 1907 (GP #100); Roosevelt to Murdo Mackenzie, July 6, 1907 (TR).

[50] *Proceedings, National Irrigation Congress, 1900*, 11; Newell to Pinchot, July 2, 1903 (GP #1937).

[51] R. H. Forbes to Pinchot, Oct. 4, 1904 (GP #2138).

to his predecessor, strongly backed the leasing measure. But the administration, its fingers once badly burned, did not try again to force it. Western cattlemen, however, continued to present similar bills to Congress and after 1909 Pinchot continued to support them publicly.[52] The cattlemen worked closely with Pinchot's organization, the National Conservation Association, during these years. They appointed representatives to public lands committees of both the National Conservation Association and the National Conservation Congress.[53] In 1912 and 1913 Pinchot, as president of the National Conservation Association, concentrated on an effort to obtain a leasing law, but, as before, representatives of settlers and homestead groups thwarted his effort.[54] When it appeared that a leasing bill might succeed in 1914, the homesteaders offered the same compromise that they had proposed ten years before in the Kincaid Act. Representative Edward T. Taylor of Colorado introduced a 640-acre stock-raising homestead bill for the entire West. Cattlemen, in turn, arguing that it would not solve their problems, fought the measure.[55] Again they lost. Congress passed the 640-acre bill in 1916, but did not approve the leasing proposition until 1934.

[52] For example, see Pinchot to Dwight B. Heard, Apr. 26, 1909 (GP #120).

[53] In 1910 the Second National Conservation Congress organized a standing committee on the public lands composed of four members, two of whom were prominent Western cattlemen working for a leasing program: Dwight B. Heard of Phoenix, Arizona, and Murdo Mackenzie of Trinidad, Colorado. *Proceedings, National Conservation Congress, 1910*, iv.

[54] See correspondence in the files of the National Conservation Association in the Pinchot MSS, under headings, "American National Livestock Association," "Dwight B. Heard," and "J. W. Tomlinson," for 1912 and 1913.

[55] J. W. Tomlinson to Harry A. Slattery, Oct. 11, 1916 (GP #1834).

Chapter V

The Public Land Question

Contacts with the West, which federal hydrographers had first established in the late 1880's, gradually drew government scientists into an increasing variety of resource problems. As the Geological Survey, the Bureau of Forestry and the Reclamation Service became involved first in irrigation, then in forest management, and finally in grazing, they began to look upon Western land problems as a whole. After one of his many Western trips, Pinchot expressed this idea:

Because of the attention directed to forestry and irrigation a new conception of public land questions, or rather of the public land question as a single problem, has meantime been coming rapidly forward, and the vital importance of it to the Nation as a whole is growing into full recognition. We are beginning to see the inter-dependence of its various parts, such as irrigation, forestry, grazing on the public lands, and the general problem of the best use of every part of the public domain, and that knowledge is becoming a principle of action, with the conception of permanent settlement at its base.[1]

These leaders were especially enthusiastic about the possibilities of vast economic growth in the West if the federal government planned the development of its resources on a large scale. By 1906 Pinchot, Newell, and other officials had formulated comprehensive land management concepts which during the remainder of Roosevelt's presidency they tried to apply to the public domain.

[1] MSS attached to letter from Pinchot to Robert Underwood Johnson, Mar. 27, 1904 (GP #89).

Elements of Scientific Land Management

This program required, first of all, a thorough revision of the public land laws, which Congress had passed originally to promote rapid disposal to private individuals rather than to aid in systematic development. To some resources no specific laws applied; for others, existing law hampered rather than promoted efficient growth. In 1873, for example, Congress passed a coal land act which provided for such a limited maximum acreage per entry that it prevented larger and more efficient coal development.[2] Laws restricting land entry to 160 or 320 acres hardly sufficed for grazing which required as much for each head of cattle. The ease with which one could evade the law, moreover, nullified any attempt to adapt a particular statute to a specific goal in resource management. Only the entryman supplied the facts concerning the nature of the land; federal agents reviewed merely a handful of applications to determine if they were valid. Anyone, therefore, could acquire lands best suited for one purpose under a variety of laws. This lack of system appalled those who were enthusiastic about more efficient growth; new management plans required major legislative reforms. "The possibilities of a wiser system of land laws," declared Pinchot, "grow to almost boundless dimensions."[3]

In the fall of 1903 President Roosevelt appointed a Public Lands Commission, "to report at the earliest practicable moment upon the condition, operation and effect of the present land laws, and on the use, condition, disposal, and settlement of the public lands."[4] Public and private groups supporting this investigation hoped that the Commission's report would persuade Congress to close the loopholes

[2] This act of March 3, 1873, limited entry to 160 acres for individuals and 320 acres for associations. In 1904 Congress increased the maximum for associations to 640 acres. These low acreage limits only placed more pressure on prospective coal operators to evade the law. Roosevelt to Ethan A. Hitchcock, Dec. 13, 1906 (TR); 59th Congress, 2nd Session, *S. Doc. 141*, 2.

[3] Pinchot to James L. Houghteling, Sept. 1, 1904 (GP #83).

[4] 58th Congress, 2nd Session, *House Doc. 1*, xxvi. J. D. Whelpley, a news correspondent, proposed the Commission to Pinchot in 1903. See Whelpley to Pinchot, Sept. 10, 1903 (GP #81); Pinchot to Newell, Sept. 18, 1903 (GP #80).

which permitted illegal entry and especially to prevent use of the homestead laws to acquire nonagricultural resources.[5] Many laws designed to aid settlers, they argued, benefited speculators and natural resource "sharks" more than homesteaders. For example, through false entries many "settlers" homesteaded valuable timberlands in small acreages, only to transfer them immediately to a single owner. The Commission repeated these contentions and recommended that Congress repeal or modify the offensive laws.[6]

To Pinchot and Newell the Public Lands Commission was to aid not only in plugging loopholes, but also in devising new laws to promote orderly development. They were most concerned, not with who acquired the public lands, but with the way in which the process of acquisition aided or hampered wise resource use. The Commission, for example, investigated the charge that speculators used the Desert Land Act to interfere with the federal reclamation program[7] and recommended that Congress modify the Act to solve this problem. Or again, giving special attention to grazing, the Commission supported a compromise between the homesteaders

[5] George H. Maxwell brought his entire propaganda organization, including the transcontinental railroads and the national manufacturing and commercial organizations, behind the Commission's work to obtain the repeal of the Desert Land Act, the commutation clause of the Homestead Act, and the Timber Culture Act. For an example of their argument that speculators were using these laws fraudulently and preventing homesteading, see Paris Gibson, "Public Domain for Home Seekers," in *Maxwell's Talisman* (Jan. 1903), 2, 3. Whether or not their views as to fraud were correct, these business organizations were primarily interested in a wide settlement and consequent economic development which would increase their trade.

[6] 58th Congress, 2nd Session, *Sen. Doc. 189, Report of the Public Lands Commission* contains the Commission's recommendations, and its account as to the condition of the public lands and the effectiveness of the public land laws. Historians have not yet gathered reliable data to test the conclusions of either the Commission or others as to the use of the land laws. See Thomas LeDuc, "The Disposal of the Public Domain on the Trans-Mississippi Plains: Some Opportunities for Investigation," *Agricultural History* (October 1950), 24, 199–204.

[7] Federal irrigation groups had complained widely of this use of the Desert Land Act. See William E. Smythe, "Robbing the People's Estate," *Maxwell's Talisman* (Dec. 1902), 2, 3–5.

and the cattlemen which Frederick V. Coville devised. Grazing lands, the Commission argued, should be classified in two groups, those which might support more intensive farming and those that could not. The former should be leased from year to year and be subject to homestead entry at the end of each grazing season. Lands which seemed to have no agricultural value, however, could be placed under ten-year permits, at the end of which time they could be reclassified into the first category.[8]

The work of the Public Lands Commission reflected the concern of the Roosevelt administration for a more orderly and planned approach to the public lands. The Commission never officially completed its work, but continued as an informal group of administration leaders who backed a new set of land management principles. The most obvious of these principles was public ownership. The old practice of disposing of nonagricultural lands to private owners, Pinchot and others argued, must give way to public ownership and public management. To the general public, the goal of public ownership and the controversy over public or private resource management was the crucial conservation issue. But, to the officials of the Roosevelt administration, public ownership was merely a means to an end; it alone would permit rational development. The significance of the new public lands program, therefore, lay, not in the method of public ownership, but in the objective of efficient, maximum development.[9]

Efficient land management also required an exact knowledge and careful classification of resources. The administration annually plunged into a fight with Congress to increase the meager appropriations for the Geological Survey which in 1877 had inherited the four military and civilian surveys in the West begun after the Civil War, and in later years had struggled to carry forward a resource

[8] Public Lands Commission, *Report,* xviii–xx.

[9] Major works on the public lands—those by Ise, Robbins, and Peffer—stress the simple problem of public ownership. In evaluating Ise's book, *United States Forest Policy,* Philip P. Wells, former law officer of the Forest Service, criticized it for describing the conservation movement primarily as a protest against "land grabbing." See Wells to Pinchot, Jan. 3, 1919 (GP#1676).

mapping program.[10] Using this information, the Roosevelt leaders classified the public lands so as to determine which areas should be reserved for particular uses. As the Director of the Geological Survey explained, "In practice land classification means simply the determination of highest use. . . . With the different values of the land made known by adequate examination, the highest use can be determined and, in so far as the statutes are in accord with economic law, the highest use can be assumed."[11] As early as 1878 John Wesley Powell had proposed that mineral, timber, coal, irrigable, and grazing lands be classified, and prior to 1905 some progress had been made toward that end. The Forest Reserve Act of 1891 set aside forest land for the specific uses of forest production and watershed protection. Since the forests included range land as well, the Forest Service itself divided or classified areas within the reserves as watershed, timber production, grazing, or wildlife areas. The grazing proposal of the Public Lands Commission involved a similar procedure by distinguishing between farming and grazing lands on the public domain.

After 1905 the Roosevelt administration began a more systematic program of classifying all public domain resources, including water power sites, and coal, oil, and phosphate lands. To guard against entry during the period of investigation, Secretary of the Interior James R. Garfield temporarily withdrew the lands under scrutiny from private sale. The Geological Survey examined the withdrawn areas and submitted its data to Department of the Interior officials who then classified each area as most valuable for a particular use. The work increased rapidly, and in December 1908 the Department organized a special Land Classification Board in the Geological Survey to carry it forward. Once the Department had classified the land, it restored to the public domain those agricultural areas not containing other valuable resources.

Who would be permitted to use these resources? Here lay the

[10] A brief account of these surveys is in William C. Darrah, *Powell of the Colorado* (Princeton, 1951), ch. 15.

[11] *Annual Report of the U. S. Geological Survey* (hereafter cited as *Ann. Rept. USGS*), *1912*, 21.

thorniest problem of all. Should irrigation development take precedence over water power? Should commercial grazing have a higher priority than game hunting? The administration never set down a definite code but did assume a rough system of priorities in attempting to resolve specific use conflicts. In the national forests Pinchot granted top priority to domestic use of water, followed by irrigation and power; rights-of-way over the public domain, as well as over the forests, he argued, should be granted in that order. On agricultural lands homesteading should precede grazing; he backed the flexible leasing plan for the public domain and granted first use of forest grazing lands to farmers with small herds of livestock.[12] The conflict between recreation and commercial use Pinchot found to be extremely hazardous to resolve, but he firmly argued that commercial use of the public lands should precede their use for recreation. Reservoirs for municipal supply of water power, for example, should be permitted in the national parks.

Efficient development required carefully regulated conditions of use as well as a scale of priorities. Grazing, for example, should not exceed the carrying capacity of the land. By issuing permits to use the public lands, the federal government could control the conditions of use. These permits, Pinchot argued, should run for only a limited period of time, so that if an area later could be developed for a more valuable use, it could be reclassified. Permits should prevent speculation by requiring prompt development. Moreover, the administration believed, public land users should pay for their privilege a fee approximately equal to that paid to owners of private lands. The administration applied these basic conditions—a limited permit, prompt use, and a user fee—to all resources on the public lands.[13]

Finally, the new land management entailed administrative innovations. Experts rather than politically appointed officials, for example, should take charge of the program. Pinchot had long emphasized both scientific training for foresters and the use of civil service examinations to select them for government work. He hoped

[12] Roosevelt to Secretary of Agriculture James Wilson, Dec. 21, 1905 (TR).

[13] Never expressed in any one concise statement, these principles were implied in many specific actions over a period of years.

to establish schools to train irrigation and land management experts, as he had for forestry, but these plans fell through.[14] Nevertheless, the Roosevelt administration constantly increased the number of trained foresters, range specialists, and geologists in its public lands program.

Public land administration, moreover, should be more integrated; Pinchot argued that all public land questions, being closely related, should come under the supervision of a single federal department. This view sprang from the interrelatedness of resources themselves, and from a desire to avoid the interdepartmental conflicts which resulted when competing resource users played one agency against another for their own advantage. As early as 1903 Pinchot brought this problem to the attention of the Committee on Scientific Methods which Roosevelt had appointed to investigate the possibilities of improving the efficiency of federal scientific work.[15] Following Pinchot's lead, the Committee recommended that the Geological Survey, the General Land Office, the Office of Indian Affairs, and all national park and forest administration be consolidated in the Department of Agriculture.[16] Congress approved only the transfer of the forest reserves, but Pinchot continued to stress the need for greater coordination.[17]

Appointment of James R. Garfield as Secretary of the Interior in 1907 reduced Pinchot's anxiety over administrative coordination. The Chief Forester had worried especially about the lack of interest that the former Secretary, Ethan A. Hitchcock, displayed in the new approach to land management. Hitchcock seemed to look upon the public lands either as areas to be preserved from use or distributed to settlers. He resisted the move to permit livestock in the forests and to lease the public domain ranges. He opposed commercial use of the national parks. Between 1903 and 1907 Pinchot

Pinchot's records

[14] Pinchot to Archer M. Huntingdon, June 20, 1905 (GP #89).

[15] Pinchot to Joseph A. Holmes, June 23, 1903 (GP #1936).

[16] "Report of the Committee on Scientific Methods," nd, MSS in GP #1937.

[17] See, for example, *Minutes of the Meetings of the Inland Waterways Commission*, May 14, 1907 (GP #2136); Pinchot to Irving Fisher, May 6, 1908 (GP #1936).

served as Roosevelt's chief adviser on public lands questions, but as long as Hitchcock was Secretary, the Chief Forester could not implement those plans as he thought best. He supported the transfer of Interior Department agencies to the Department of Agriculture largely to overcome Hitchcock's resistance to the new public lands policies.

Garfield, on the other hand, agreed with Pinchot's views, and was offered the post of Secretary primarily because of this fact. The son of former President James A. Garfield, and since 1888 a Cleveland lawyer, the new Secretary attracted the attention of Theodore Roosevelt early in his administration. In 1902–03 he served on the United States Civil Service Commission, and from 1903 to 1907 was Commissioner of the new Bureau of Corporations. Garfield had had little experience with conservation problems, but his views on antitrust questions embodied the same spirit of planned, efficient development which the administration's resource leaders revealed. Garfield thoroughly agreed with Roosevelt that big business was efficient business, and that the Sherman Antitrust Act looked backward rather than forward. This approach, so far as Pinchot and Roosevelt were concerned, admirably qualified Garfield to administer the Department of the Interior. From the middle of 1906 on Garfield and Pinchot worked together to establish the outlines of a new Interior Department, and the beginning of the new Secretary's term, March 1907, marked the complete victory within the administration for the program of scientific land management.

Water Power on the Public Lands

The Roosevelt administration set out to apply its new policies first to water power development on the public lands. Success came slowly, however, and in the face of bitter opposition from many in the West. In the long-drawn-out battle to establish a power program, conservation leaders tackled a variety of management problems: resource classification, curbs on speculation, a user fee, and establishment of the principle of public ownership. To each of these they applied general views which they had formulated in the years before 1907.

The history of water power conservation goes back to the 1890's when Congress, in several acts, provided for right-of-way grants across public lands for irrigation ditches and reservoirs. A law of February 15, 1901 codified these acts and delegated to the Secretary of the Interior power to grant rights-of-way "under general regulations to be fixed by him," and revocable at the will of the Secretary. Under this act the Department of the Interior issued permits for water power reservoirs and generating facilities as well as irrigation works, but it made no attempt to establish any public regulation or control.[18]

After Congress transferred the reserves to the Department of Agriculture, the Secretaries of the two Departments formally agreed that the Forest Service would grant temporary permits, but that the Department of the Interior would rule on permanent easements.[19] Since the Act of 1901 granted only temporary permits, its administration, so far as the national forests were concerned, after 1905, rested with the Department of Agriculture. This agreement was of the utmost importance for the water power industry in the national forests. For the Act of 1901, argued Pinchot and the Forest Service, delegated to the administration the responsibility of attaching to all permits specific conditions which it felt were essential to protect the public interest. Under this Act the Forest Service developed an extensive regulation of the water power industry within the national forests.

The Forest Service drew up a standard permit agreement which it required all users to sign. It contained the following terms: an easement of definite tenure; a time limit for the easement, determined by the Secretary of Agriculture to suit the needs and magni-

[18] Philip P. Wells, "Personal History," MSS in GP #1671. Wells was chief legal officer for the Forest Service from March 3, 1907 to the fall of 1909; his main task was to work out the water power policy for the forest reserves.

[19] *Decisions of the Department of the Interior and General Land Office*, v. 33, p. 609 (Washington 1905). Pinchot described this agreement as a mutual interpretation of the transfer act by the two departments that "all grants or privileges within Forest Reserves which did not affect the title to the land or cloud the fee were under the jurisdiction of Agriculture. . . . All those which did remained under Interior." Pinchot, *Breaking New Ground*, 334.

tude of each project; a requirement that construction be completed in a definite and reasonable time to prevent speculation; and an annual charge of an amount the Secretary might deem proper and change from year to year as circumstances might warrant. In 1906 Pinchot included these principles in a special law granting to the Edison Electric Company a water-power permit in a California national forest.[20] This act, he hoped, would pave the way for a uniform law covering rights-of-way and privileges on all federally owned land.[21]

Private power corporations strongly opposed these innovations. The regulations, they argued, should grant perpetual leases and exclude all charges.[22] The federal government, they continued, could not constitutionally control running water on the public lands because water rights were derived from state law. Since the state had granted them a free water right, the federal government was morally obliged to grant a free land tenure.[23] Pinchot submitted the dispute to Attorney-General Charles J. Bonaparte, who on October 5, 1907 sustained the Forest Service and even intimated that states could not assume control over the use of water in the national forests. In August 1908 Judge Robert E. Lewis of the United States Circuit Court of Colorado also upheld the permits.

The power corporations tried to persuade Congress to pass a measure granting perpetual leases with a nominal charge of only $2.50 an acre. Senator Crane of Massachusetts and Representative Mondell of Wyoming introduced such a bill in the winter of 1907–08, as did several other lawmakers.[24] Roundly condemned in a

[20] For a description of the provisions of this act and its implications, see *Forestry and Irrigation* (July 1906), 12, 325.

[21] A complete account of the water-power permit provisions is in Wells, "Personal History" (GP #1671).

[22] Richard M. Saltonstall to Roosevelt, Dec. 17, 1907 (GP #1666).

[23] Wells, "Personal History" (GP #1671), describes the legal arguments used on both sides of the issue.

[24] Saltonstall to Roosevelt, Dec. 17, 1907; Saltonstall to Pinchot, Feb. 1, 1908 (both in GP #1666). Saltonstall was a member of a Boston law firm handling the legal work for the Stanislaus Power Development Company, which planned to develop water power in the Stanislaus National Forest in California.

message to Congress on February 26, 1908[25] by President Roosevelt, these proposals failed to pass. Checked here, the companies requested a conference with the administration with a view to reaching an agreement. In the early spring of 1908, attorneys, bankers, and hydroelectric managers from Boston, New York, Washington, and Los Angeles met with federal officials in Washington. Pinchot agreed to one minor compromise, that a definite fifty-year permit, irrevocable except for violation of its conditions, replace the section of the Act of 1901 under which the Secretary could revoke a permit at his discretion. This failed to satisfy the power companies, and the conference ended in a deadlock. The companies held out for the Crane-Mondell provisions, and Pinchot, Garfield, Newell, and their law officers demanded a fifty-year maximum and an annual charge.[26]

During the summer of 1908 Pinchot discovered that some corporations had evaded the permit system by entering power sites under the mineral land laws.[27] The Forest Management Act of 1897 permitted mineral entry anywhere within the national forests. The mineral laws, therefore, could serve conveniently as a device to acquire valuable nonmineral lands. This loophole constantly annoyed the Forest Service. Frequently it could not use strategically located administrative sites because private parties had already taken them up as mineral land. To check this practice, the Secretary of the Interior, in an arrangement with the Chief Forester, withdrew from all forms of entry those sites which the Forest Service wished to use for administrative purposes. Pinchot used the same device to forestall water power entries under the mineral laws. At his request, Secretary Garfield withdrew from all forms of entry 100 to 150 acres within each water-power site under the pretext that they were valuable locations for ranger stations.[28] Later, after he had formulated the doctrine of executive "supervisory power"—that the

[25] 60th Congress, 1st Session, *Sen. Doc. 325, Preliminary Report of the Inland Waterways Commission*, v.

[26] Wells describes this conference and its results in "Personal History" (GP #1671).

[27] See untitled MSS about this problem by Pinchot in GP #2139.

[28] *Ibid.* See also Rose M. Stahl, "The Ballinger-Pinchot Controversy," *Smith College Studies in History* (Jan. 1926). XI (2), 84–86.

President could constitutionally withdraw land for any purpose that would aid in the execution of the land laws—Secretary Garfield again withdrew these lands as straight water-power sites. This time, however, he withdrew them from all forms of entry *except* under the Right-of-Way Act of 1901, and, therefore, under the Forest Service permit system.[29] Contrary to the arguments of many of his contemporaries, and of some historians since then, these withdrawals did not bar present use in favor of future development. They were, in effect, a classification of land for immediate, though specific, use.

Pinchot tried to apply his power policy not only to new installations, but also to those companies which had obtained permits from the Secretary of the Interior prior to the transfer of the reserves. Some which had already constructed plants refused to comply and Pinchot requested the Attorney-General to enjoin or eject them. These suits, instituted under the Roosevelt administration, dragged on until President Taft's Attorney-General, George Wickersham, postponed the proceedings. But Walter L. Fisher, who became Taft's Secretary of the Interior in March 1911, persuaded the Attorney-General to take them up in the spring of 1913. Finally in 1917 the Supreme Court upheld the legality of the permit and the charge.[30] In other instances power companies had undertaken no development under permits received from the Secretary of the Interior prior to 1905. Arguing that the permittees held the sites merely for speculation, the Forest Service initiated action to revoke their grants. On March 2, 1909, two days before his term of office ended, Secretary Garfield signed the revocation orders, an action which the power companies bitterly condemned.[31]

[29] Wells to Arthur W. Page, Aug. 15, 1910 (GP #1670); Pinchot to S. A. Kean, Aug., nd, 1910 (GP #1819); Wells, "Personal History" (GP #1671); Wells to Thomas J. Walsh, nd, (GP #1666); 61st Congress, 3rd Session, *Sen. Doc. #719, Investigations of the Department of the Interior and the Bureau of Forestry*, v. 2, 276.

[30] *Utah Power and Light Co.* v. *United States*, 243 U. S. 389 (1917).

[31] Philip P. Wells to editor, *The Outlook*, Nov. 8, 1910 (GP #1819); this letter was written in reply to an article which J. R. McKee wrote in the October 1910 issue of the *North American Review*, singling out the revocation orders for attack.

Checked by the administrative agencies and the courts, for over a decade the power companies continued to press for favorable legislation. They also accepted the Forest Service permit system, and as a result hydroelectric development in the national forests went forward rapidly. Between 1909 and 1916 installed capacity of plants in the national forests increased from 250,000 to 738,092 horsepower, while the Forest Service by 1916 had granted permits for 810,300 more. Of the total capacity in 1916, 222,030 horsepower represented plants with reservoirs alone in national forests; and 536,062 horsepower, plants actually erected there. Under the permit system, power facilities in the national forests had increased to a point where, by 1916, they constituted 42 per cent of the total developed power in the Western states. By this time, hydroelectric power production in the West exceeded demand and a market shortage was retarding construction. During 1915, for example, peak loads reached only 61·1 per cent, and total output only 34 per cent of the hydroelectric capacity in the Western states. At the same time a potential of almost 300,000 horsepower lay undeveloped at sites owned by private corporations. It could hardly be said that Forest Service regulations retarded the growth of hydroelectric power.[32]

Prior to March 4, 1907, when Secretary Hitchcock left office, the Department of the Interior declined to follow Pinchot's interpretation of the Act of 1901 and continued to grant perpetual and unrestricted rights-of-way over the public domain. Charges for permits under that Act, the departmental lawyers argued, were illegal, and Attorney-General Bonaparte's ruling to the contrary applied only to the national forests and not to the public domain. James R. Garfield, the new Secretary of the Interior, hoped to reverse this policy and apply the Forest Service permit system to all the public lands.[33] In fact, Roosevelt appointed Garfield to the Interior post partly for that

[32] This data is taken from Oscar C. Merrill, Chief Engineer of the U. S. Forest Service, to Philip P. Wells, Mar. 9, 1916 (GP #1665), and from Oscar C. Merrill MSS, "Water Power Development," dated Jan. 1, 1916 in GP #1665.

[33] Pinchot to S. A. Kean, Aug., nd, 1910 (GP #1819).

very purpose. To work out the legal basis for the change, the President promoted Pinchot's law officer, George Woodruff, to Assistant Attorney-General, assigning him to the Department of the Interior. Woodruff, in complete agreement with Pinchot's power views, became the guiding genius behind the new Interior policies.

From Pinchot's experience with the national forests, Garfield realized that water power corporations could easily circumvent a permit system by entering sites under other public land laws. To prevent this, the new Secretary withdrew water power sites from all forms of entry *except* under the Right-of-Way Act of 1901. By this action he hoped to segregate all lands chiefly valuable for water power and to permit their entry under water power laws alone. The orders did not withdraw these sites from all forms of use, as many then believed and others since have argued, but reserved them for a specific use.[34] The sites for "withdrawals" were selected on the recommendation of the Bureau of Reclamation and totaled, nominally, 3,928,780 acres. This figure was misleading, for the areas, only roughly mapped out, contained much privately owned land within their boundaries. The "withdrawal" orders, however, expressly *excluded* all "lands the title to which has passed from the United States."[35]

Secretary Garfield was able to go no farther with his plans. President Taft thwarted his attempt to apply the Forest Service permit system to public domain water power sites by appointing Richard A. Ballinger as the new Secretary of the Interior. Ballinger not only declined to carry forward Garfield's policies, but, in fact, tried to reverse them. He refused to grant any water power permits at all, and for the first time, during his two years as Secretary, public domain water power sites were locked up and all new development ceased.

Walter L. Fisher, on March 4, 1911, succeeded Ballinger as Secretary of the Interior, and continued Garfield's policy where his

[34] Wells to Arthur W. Page, Aug. 15, 1910 (GP #1670); see also Wells, "Personal History" (GP #1671).

[35] See MSS by A. P. Davis, Chief Engineer of the Bureau of Reclamation, who supplied the data for the withdrawals, in GP #120.

predecessor had halted it two years before. For almost two years Fisher worked out the details of a permit system. To help formulate his policy he persuaded Philip P. Wells, Pinchot's former law officer, to come to the Department of the Interior. He kept in close touch with private power companies. On April 8, 1911, after Fisher had addressed the Power Transmission Section of the National Electric Light Association, that body appointed a committee to confer with the Secretary on water power problems.[36] On August 24, 1912, Fisher issued the Department's first water power regulations. After further conference with the private corporations, he issued revised regulations on March 1, 1913. The Forest Service cooperated by modifying its permit so as to conform more closely with the Interior Department policy. The new permit called for a rental adjustment every ten years, rate and service regulation by state public utility commissions, prohibition of capitalization of the permit, and public recapture, in case of a violation, on fair terms.[37] Since the General Land Office still regarded the permit system as illegal, Fisher gave to the Geological Survey the task of administering it. Thus, Secretary Fisher finally carried out Garfield's objective, and once more opened to development the water power sites which Ballinger had withdrawn from use.

Throughout these years those who worked to devise an administrative permit system under the Act of 1901, sought to modify that Act in one important respect: to guarantee that permits would run for a minimum length of time. The Act had declared all permits revocable at the will of the Secretary of the Interior. "While it is the purpose of the Department in the administration of the waterpower on the National Forests to take no advantage of such a situation," declared the Chief Engineer of the Forest Service, "it is realized that such a purpose is not a satisfactory safeguard where

[36] Overton W. Price to Pinchot, Apr. 4, 1911 (GP #1817); *American Conservation* (Apr. 1911), 1, 109; Philip P. Wells to Henry A. Doherty, Apr. 13, 1911 (GP #1665); Price to Pinchot, Apr. 15, 1911 (GP #1817); *New York Evening Post*, Nov. 25, 1912 (GP #1665).

[37] *Ann. Rept., USGS, 1912*, 88; *1913*, 136, 157–158.

millions of investment are concerned."[38] In 1907 the Forest Service proposed that Congress guarantee permits with minimum duration of 50 years, but the power companies, holding out for more favorable terms, blocked the change. Later efforts also met defeat. However, in 1911 Congress did authorize rights-of-way for transmission lines across the public lands under 50-year minimum grants.[39] Under this law Secretary Fisher negotiated successfully with the Montana Power Company a permit which included compensation, publicity of operations and strict federal regulation, as well as the 50-year minimum franchise.[40]

With these innovations under Secretary Fisher, the Roosevelt water power policy was fully applied to the public lands. Private power corporations, accepting the new policies, rapidly developed hydroelectric projects in the West. They also pressed for more favorable legislation. While they agreed to accept elaborate federal regulations and became more amenable to the 50-year limited grant, they steadfastly refused to approve the permit charge. Their proposals in Congress met determined opposition from Gifford Pinchot, at this time president of the National Conservation Association, who continued to demand that new legislation include the right to impose a fee. This problem became so closely identified with the conservation movement that Congress frequently referred to the water power fee as a "conservation charge." For over a decade these two contending groups were deadlocked; Pinchot's main support lay in the House, while the Senate usually agreed with the power companies. The struggle finally ended with the Water Power Act of 1920, which established the principle of public regulation of hydroelectric power, but provided for much less federal revenue than the Roosevelt conservationists had desired.[41]

[38] Oscar C. Merrill to Louis B. Stillwell, Dec. 23, 1910 (GP #1665).

[39] Act of March 4, 1911 (36 *Stat.* 1253). Regulations under this act were drawn up Jan. 6, 1913; see *Ann. Rept. USGS, 1913*, 158.

[40] *Electric Railway Journal* (Jan. 11, 1913, 41, no page, attached to letter John D. Ryan to Philip P. Wells, Jan. 13, 1913 (GP #1666).

[41] For the background of the Water Power Act of 1920 see Jerome Kerwin, *Federal Water-Power Legislation* (New York, 1926), 170–262. See also the

The New Coal Policy

The new public land mineral policy, especially as applied to coal, provoked as much controversy as did water power. The coal problem first came to the serious attention of the administration during the summer of 1906 when it learned that "prospectors" had filed coal claims of slight coal value on national forest lands in order to acquire the standing timber on the surface. Further investigation aroused the suspicion that private parties also frequently acquired valuable coal deposits by fraud under the homestead laws, and held them for speculation.[42]

Pinchot and Garfield were convinced that existing statutes encouraged these practices. For example, the homestead laws granted the settler title, not only to the surface area, but to the coal deposits beneath as well. Through fraudulent use of these laws, therefore, a speculator could acquire larger coal deposits than the coal laws permitted. On the other hand, in the national forests lax administration permitted enterprising land sharks to acquire forest and other resources through the coal laws. The Forest Management Act of 1897 permitted mineral entries in the forests, and the General Land Office, which granted permanent easements in such cases, conducted on-the-spot claim inspections only if they were contested. The coal land laws, moreover, provided for maximum entries hardly sufficient to establish a profitable mine. At first the law limited entry to 160 acres, but in 1904 Congress directed that four such entries could be consolidated for a joint enterprise. Yet this also did not suffice, and coal companies continued to evade the law to secure larger areas.[43]

Low prices, far out of line with coal prices charged by private landowners or the Western states, the administration argued, also stimulated false entries and speculative holdings. The coal act of

---

files of the National Conservation Association, 1910–1920, located in the Pinchot MSS.

[42] Philip P. Wells describes the coal problem in an untitled MSS on conservation history in GP #1876.

[43] Pinchot, *Breaking New Ground*, 396; 59th Congress, 2nd Session, *Sen. Doc. 141*, 2.

1873 had provided for a price "not less than" $20 per acre for coal lands within fifteen miles of a railroad, and "not less than" $10 for those beyond fifteen miles. Prior to 1906 the General Land Office had interpretated "not less than" to mean "not more than" and had charged a flat price of $20 and $10 per acre. It conducted no investigations to determine the value of the coal itself, and resisted all attempts to charge prices more in line with the worth of the lands. Due to the wide gap between the federal price and private coal land values in the West, the speculator could make a rapid capital gain by acquiring public land and selling it for a much higher price.[44]

Roosevelt, Pinchot, and Garfield set out to change these policies by requiring classification of all coal land, development of coal under coal law alone, and disposition of coal lands at prices more in accord with their market value. Relying upon the advice of Charles D. Walcott, director of the Geological Survey, the administration in the summer of 1906 drew up the necessary legislation and took initial departmental steps to further its new policy.[45] On June 29, 1906 Roosevelt officially requested the Secretary of the Interior to determine which coal land the administration should investigate to determine its value.[46] He then temporarily withdrew this land from entry—some 50,000,000 acres—in order to examine and classify it, and establish its value. These withdrawals continued during succeeding years, and by July 1, 1916 totaled 140,533,745 acres.[47]

The Geological Survey then undertook the enormous task of determining the exact quantity and quality of coal on the public lands.[48] Its work proceeded rapidly. By March 1907, 28,000,000 acres containing no coal of sufficient value for permanent classification as coal lands were restored to entry.[49] The Land Classification Board, organized in December 1908, continued this work in more systematic fashion. By July 1, 1916 the Board had examined a total

[44] Van Hise, *Conservation of Natural Resources*, 35.

[45] Roosevelt to Ethan A. Hitchcock, Dec. 15, 1906 (TR).

[46] Roosevelt to Hitchcock, June 29, 1906 (TR).

[47] This figure is compiled from the annual reports of the U. S. Geological Survey, 1908–1916.

[48] Roosevelt to James R. Garfield, Mar. 12, 1907 (TR).

[49] *Ann. Rept. USGS, 1907*, 4.

of 94,597,971 acres, of which it had classified and restored to entry 73,602,780 acres as non-coal-bearing land. By that date, therefore, 20,995,191 acres remained classified as coal lands.[50]

These classified lands the administration opened to entry under the coal land laws alone. Pinchot and Garfield particularly hoped to end the practice of acquiring valuable subsurface resources by taking up homestead claims. Since the law made no provision for entering agricultural lands independent of the coal deposits beneath, no homesteader could acquire surface rights alone within the classified areas even if he had no interest in the mineral deposits. The Land Classification Board received many requests that withdrawn lands be reclassified as non-coal in order to facilitate their agricultural entry. Some of these applications the Board granted, but only where the coal involved, because of its insufficient value, could not be developed economically.[51] In 1910 and 1911 Congress solved this problem when it approved two bills sponsored by Representative Mondell of Wyoming, which permitted new agricultural entries on withdrawn coal lands if the entryman would agree to leave all subsurface rights with the federal government.[52]

The plan to increase the price of coal lands became the most controversial of all the new coal policies. The administration proposed that the federal government retain ownership of the coal lands, lease them for a per-ton royalty, and apply regulations necessary to protect the public interest. Classification had required only administrative action, but Congress would have to approve a leasing program. Roosevelt first proposed the coal-leasing measure in his message to Congress on December 17, 1906.[53] George Woodruff, then law officer of the Forest Service, and Joseph A. Holmes of the

[50] These figures are compiled from the annual reports of the U. S. Geological Survey, 1908–1916.

[51] *Ann. Rept. USGS, 1911*, 66; *1912*, 94; *1913*, 149; *1914*, 134; *1915*, 159; *1916*, 154.

[52] Acts of March 3, 1909 and June 22, 1910. For this problem see 61st Congress, 2nd Session, *Sen. Rept. 703, Agricultural Entries on Coal Lands.*

[53] 59th Congress, 2nd Session, *Sen. Doc. 141*, 2; Roosevelt to Hitchcock, Dec. 13, 1906 (TR).

Geological Survey drafted an administration bill.[54] Senator Knute Nelson of Minnesota introduced the measure, but the Senate Committee to which it was referred took no action. The majority of the House Public Lands Committee approved the proposed law because it would permit leasing while not excluding sale, thus providing an opportunity to test the two systems side by side. Although the Committee's minority condemned the measure as "dangerously socialistic, paternalistic, and centralizing in character," the majority favorably reported it to the House.[55]

Roosevelt personally took up the fight to pass the leasing bill.[56] "The nation," he argued in a special message to Congress on February 13, 1907, "should retain its title to its fuel resources, and its right to supervise their development in the interest of the public as a whole."[57] Even before this message, however, conferences with Western congressmen had convinced the President that the leasing measure could not pass over their determined opposition. He, therefore, shifted his ground, and proposed a bill which established leasing as a principle, but did not go into the details of leasing itself. These, he hoped, could be worked out with the coal operators at a later date, and be written into the law as an amendment. At first he planned to impose very liberal leasing requirements in order to meet Western opposition, but once the principle was established the administration could modify the terms in line with its basic objectives.[58]

Roosevelt's measure, however, met stubborn resistance from one of the Senate leaders in mineral law reform, Robert LaFollette of Wisconsin. As early as June 1906, LaFollette had introduced a concurrent resolution to authorize the withdrawal of coal, oil, and lignite lands, and he continued to play a leading role in the movement for new mineral land laws. He hoped to fix by law the matters

[54] Philip P. Wells, MSS on conservation history in GP #1876.

[55] 59th Congress, 2nd Session, *House Rept. 7643*, 7, 13.

[56] Roosevelt to Senator Spooner, Jan. 11, 1907; Roosevelt to Senator Hansbrough, Jan. 29, 1907; Roosevelt to Lacey, Jan. 22, 1907; Roosevelt to LaFollette, Jan. 23, Feb. 5, Feb. 19, 1907 (all in TR).

[57] 59th Congress, 2nd Session, *Sen. Doc. 310*, 1.

[58] Roosevelt to LaFollette, Jan. 23, Feb. 3, 1907 (TR).

which Roosevelt wished to leave to the discretion of the Secretary of the Interior. The President appealed to LaFollette, without success, to consider the substitute proposal as politically preferable. The two men remained apart not simply because they disagreed on tactics; they differed on the basic elements of public policy. Roosevelt, LaFollette feared, would by administrative action approve coal combinations which ought to be prohibited by law. The President opposed the Senator for the same reason; LaFollette's bill, he was afraid, would forbid all coal combinations. "My experience," Roosevelt wrote to the Senator on February 19, 1907, "tends to make me believe . . . that the point to be aimed at is not so much an indiscriminate forbidding of all combinations for whatever purpose, but rather a supervision which will prevent noxious combinations and will insure any combination that does take place being in the interest of the public."[59] The President's plea failed. Opposition from the West and from LaFollette were enough to kill the administration measure and to prevent its further consideration during Roosevelt's term of office.

Failing to secure congressional action, Roosevelt took administrative steps to raise the price of coal lands. The Act of 1873, he argued, implied that the Department of the Interior could fix the price of coal at its discretion. That Act provided for minimum prices, but did not establish maximums; the Secretary of the Interior, therefore, could charge a price higher than the minimum. The Geological Survey, after classification, valued the coal lands according to the quality of the deposits and their accessibility, and established prices which would aid development, while preventing speculation.[60] Up to July 1, 1909, the Survey had valued 742,573 acres of coal lands at a total of $30,488,351; this was twice as much as the minimum price of $14,142,712 for the same area provided for in the Act of 1873.[61] After the Survey had completed the valua-

[59] Roosevelt to LaFollette, Feb. 19, 1907 (TR).

[60] Walter L. Fisher to Frank W. Mondell, Aug. 17, 1912 (GP #1672).

[61] *Ann. Rept. USGS, 1909,* 44–45.

tions, the Department reopened the lands to entry and sales went forward rapidly.[62]

Despite the success of the new sales program, conservationists continued to work for a leasing measure. Late in 1907 the administration sent A. C. Veatch of the Geological Survey to Australia and New Zealand to study the leasing systems in operation there.[63] In 1910 a congressional Joint Committee investigating the activities of the Department of the Interior and the Forest Service recommended that Alaskan coal lands be leased instead of sold. For the next decade Pinchot, as president of the National Conservation Association, fought for a general mineral-leasing law.[64] In 1914 Congress enacted a coal-leasing measure for Alaska alone, and in 1920 applied this approach to all minerals on the entire public domain.[65] Thus was achieved the objective which Roosevelt first set forth in 1906.

Phosphate and Oil Lands

The administration applied its new mineral policy to phosphate, oil, and gas deposits, as well as to coal. At the suggestion of Charles Van Hise, President of the University of Wisconsin, who argued that phosphate miners were shipping abroad supplies that farmers in the United States would one day need for fertilizer, Roosevelt withdrew Western phosphate lands from entry on December 10,

[62] It has been commonly thought that until the Mineral Leasing Act of 1920, the coal deposits were "locked up for purposes of conservation. . . ." See Roy Robbins, *Our Landed Heritage* (New York, 1942), 368 for this point of view. Yet the Roosevelt administration did not intend to hold coal for future use, nor was this the result of its policy. Once the leasing proposition failed in Congress, Roosevelt immediately opened coal lands to entry, though under higher prices, and sales continued as rapidly as before. Between 1908 and 1914 there were 2372 coal land entries, comprising an area of 355,418·95 acres, and sold at an average price of $16·91 per acre. From 1900 to 1907 there had been 1285 entries, totaling 189,527·13 acres, and sold at an average price of $14·00 per acre.

[63] Roosevelt to A. C. Veatch, Sept. 19, 1907 (TR).

[64] See files of the National Conservation Association in GP MSS.

[65] Peffer, *Public Domain*, 130–132 has a brief account of these laws.

1908.[66] These withdrawals, primarily in Utah, Wyoming, Montana, and Idaho, continued from year to year; the Geological Survey examined, classified, and valued them.[67] By July 1, 1916 it had examined 6,947,897 acres of withdrawn phosphate lands and had restored to entry all but 2,506,478 acres.[68] Since these withdrawals included the surface as well as mineral deposits, Congress on July 17, 1914 applied to them the principle of separating agricultural and mineral entry on the same lands. Since few entrymen wished to take up phosphate lands, they did not become a major conservation problem.

Oil lands, on the other hand, were in such great demand that the first requests for a sounder federal policy came from oil prospectors themselves.[69] The root of the trouble lay in the familiar story that Congress had passed no laws specifically for the entry of oil lands. Prospectors entered oil deposits under the placer laws, which required that they actually discover oil before filing entry.[70] While searching for oil, which required heavy equipment and large investments, the prospector had no legal protection against the unscrupulous land shark who would enter the same land under the agricultural laws and hold out for a high, speculative price. By 1900 prospectors in California and Wyoming had requested the Commissioner of the General Land Office to withdraw lands temporarily from agricultural entry, so that they could drill with some degree of security. The Land Office complied, but, after it had granted some twenty requests covering a wide area, agricultural entrymen protested; and in 1903 and 1904 the Department began

[66] Roosevelt to Van Hise, Aug. 10, 1908; Dec. 16, 1908 (TR).

[67] *Ann. Rept. USGS, 1909*, 43; *1910*, 47; *1911*, 67; *1912*, 96–97; *1913*, 153; *1914*, 136; *1915*, 160.

[68] Figures compiled from the annual reports of the Geological Survey.

[69] The standard work on federal oil policy is John Ise, *The United States Oil Policy* (New Haven, 1926). An excellent collection of documents for the period before the First World War is Max W. Ball, *Petroleum Withdrawals and Restorations Affecting the Public Domain, U. S. Geological Survey, Bulletin 623* (Washington, 1916).

[70] A full discussion of placer law as applied to oil is in Ball, *Petroleum Withdrawals*, 27–59.

to restore the withdrawn oil lands. In 1907 Richard A. Ballinger, Commissioner of the General Land Office, canceled all but two of the remaining withdrawals.[71]

Prospectors still clamored for protection. They received far more sympathy for their problem from the new Secretary of the Interior, James R. Garfield, who took administrative action to classify oil lands and permit their entry exclusively under the mineral laws. In response to an urgent appeal from a Geological Survey field worker, in August 1907 Garfield withdrew from agricultural entry a large area of oil deposits in California.[72] After the Survey had examined and classified the area, he restored much of it to agricultural entry, but the oil lands he opened to entry under the mineral laws alone.[73] Two years later the Survey's field workers discovered that speculators were locating oil land under the guise of gypsum entries. The Director of the Survey requested the new Secretary of the Interior, Richard A. Ballinger, to permit oil locations alone, rather than general mineral locations of any kind, on land classed as oil land.[74] Ballinger declined to take this action, nor did he recommend that Congress enact a law to solve the problem.

Curiously enough, Secretary Ballinger, who frequently argued that the new conservation policies locked up resources from present use for the benefit of future generations, responded far more favorably to the suggestion that he close the oil lands to all forms of entry. In February 1908, George Otis Smith, Director of the Geological Survey, had recommended that Secretary Garfield, in order to reserve a future fuel supply for the Navy, withdraw oil lands from all mineral, as well as agricultural, entry,[75] but Garfield

[71] All the withdrawal and restoration documents are in Ball, *Petroleum Withdrawals*, 60–100.

[72] Ralph Arnold to George Otis Smith, June 20, 1907 (*Ibid.*, 102); George Woodruff to Commissioner, General Land Office, Aug. 15, 1907 (*Ibid.*, 102–103).

[73] George Otis Smith to Commissioner, General Land Office, June 17, 1908 (Ball, *Petroleum Withdrawals*, 103).

[74] Smith to Secretary of the Interior, June 4, 1909 (*Ibid.*, 118).

[75] Smith to Secretary of the Interior, Feb. 24, 1908 (Ball, *Petroleum Withdrawals*, 104).

refused. In September 1909, however, when the Director approached Ballinger with a similar request, the new Secretary responded with far greater enthusiasm. Director Smith buttressed his plea with a second argument, with which Garfield probably would have agreed, that sale of oil lands by the acre took no account of the value of the oil; it should be sold by the barrel. Ballinger took up Smith's arguments, and on his recommendation President Taft on September 27, 1909 withdrew the existing partial withdrawals from all forms of entry, *including* under mineral law.[76] This type of oil withdrawal continued for many years, until by 1916 Presidents Taft and Wilson had created over 50 oil reserves totaling 5,587,000 acres.[77]

In an Act of July 17, 1914 Congress opened the surface area of these lands to settlement under agricultural laws. Not until the Mineral Leasing Act of 1920, however, did it provide for disposition of the oil. In the meantime, the Department of the Interior did not work out a leasing system through administrative action as it had done with water power. Oil prospectors implored Secretary Fisher to do so in 1912, but he felt that he had no legal authority to act.[78] Leasing bills came up frequently in Congress, but not until May 1916 did one come out of Committee in the Senate. Gifford Pinchot and the National Conservation Association provided much of the opposition to the leasing bills in Congress. Most of them would have granted patents to those who had filed on oil lands after their withdrawal. This, argued Pinchot, would play into the hands of speculators. Finally, on February 23, 1920 Congress passed a compromise act providing that the federal government retain ownership of the land, lease mineral rights to private companies, and return the revenue to the West for irrigation and other purposes.[79]

[76] Ballinger to Taft, Sept. 17, 1909 (*Ibid.*, 134–135); Ballinger to Frank Pierce, Acting Secretary of the Interior, Sept. 26, 1909 (*Ibid.*, 135). This was a departmental, and not an executive order; therefore it was made by the Secretary and not by the President.

[77] Ball, *Petroleum Withdrawals*, 25.

[78] Walter L. Fisher to A. C. Foster, Nov. 29, 1912 (GP #1672).

[79] Ise, *The United States Oil Policy*, 324–355.

Chapter VI

Taming the Nation's Rivers

While the Roosevelt administration, as a result of its experience with forest and range problems, was expanding its public lands policies, it was also developing the principles implicit in the reclamation program into broader views of multiple-purpose river development for the entire country. Those in the forefront of the irrigation movement—Newell, Maxwell, and Newlands—played a leading role in forming this policy. They were joined by two new federal officials, WJ McGee and Marshall O. Leighton, and were aided by a growing popular demand that the federal government promote inland navigation. By 1908 the administration had come out squarely for the multiple-purpose development of the nation's rivers through a commission that would have broad powers to authorize and construct projects. Although Congress approved few of its proposals, the Roosevelt administration for the first time worked out the general principles and the specific elements of the multiple-purpose approach to river development which the New Deal put into practice over two decades later.

The Waterway Renaissance

In the years after 1895 the United States witnessed a new enthusiasm for the improvement of its navigable streams. Many communities throughout the country seemed to catch a vision of the unlimited possibilities for local economic growth which cheaper transportation could create.[1] Urban merchants and manufacturers

[1] Edward Lawrence Pross, "A History of Rivers and Harbors Appropriation Bills, 1866–1933" (PhD Thesis, Ohio State University, 1938), 139.

spearheaded this new drive.[2] After 1898 shippers faced a steady rise in railroad rates which in previous decades had fallen sharply. To combat these increased costs, they worked for legislation to strengthen the power of the Interstate Commerce Commission to regulate railroad rates, and to promote inland navigation which would bring competition into the field of transportation. The interests of merchants and manufacturers soon became merged with the larger interests of the entire community, as local and regional waterway publicity groups and newspaper editors warned that the future growth of the community itself depended on cheaper transportation. The campaign for waterway improvement, therefore, took on the character of local patriotism which congressmen were eager to follow by demanding a larger share of the federal rivers and harbors appropriation.[3]

Two broad proposals dwarfed all others. One envisaged a deep intercoastal waterway from Boston to the Rio Grande via a sea-level cross-Florida canal.[4] The other projected a deep channel, navigable by ocean-going vessels, from Chicago to the Gulf of Mexico by way of the Illinois and Mississippi rivers.[5] Other proposals, far less comprehensive, received equally enthusiastic backing. Many eagerly supported plans for deepening the Missouri and the upper Mississippi, for improving the Arkansas and the Columbia. Private capital again seriously considered the Lake Erie–Ohio River canal, by way of Pittsburgh, and sought congressional authorization to undertake it.[6] And Atlanta, Georgia interests demanded that the federal government survey a canal route from the Ohio River through the

[2] See, for example, Chicago Commercial Association, *From the Great Lakes to the Gulf of Mexico* (Chicago, 1906), 3–19.

[3] This was especially true of the demand of the Mississippi Valley for recognition. See *St. Louis Post-Dispatch*, Apr. 7, 1907 (GP #2137).

[4] 61st Congress, 1st Session, *Congressional Record*, July 27, 1909, 4609–4612.

[5] Charles H. Harvey, "The Lakes-to-the-Gulf Deep Waterway Association," *World To-day* (Jan. 1907), 12, 39–41.

[6] W. Frank McClure, "A Barge Canal Between Pittsburg and Lake Erie," *World To-day* (March 1906), 10, 323–324.

Tennessee, the Coosa, the Ocmulgee, and the Altamaha rivers to the Atlantic Coast![7]

The House Rivers and Harbors Committee, which drew up the rivers and harbors bills, and the Corps of Engineers, in charge of the construction work, did not respond favorably to this enthusiasm. In 1899 Theodore E. Burton, one of the few congressmen to master the complex details of rivers and harbors legislation, became chairman and the dominant figure of the Rivers and Harbors Committee. Burton, the son of a minister, and a graduate of Oberlin College in 1872, had practiced law in Cleveland for over a dozen years before he was elected to Congress in 1889. He served in the House until 1909, with a short intermission of four years between 1891 and 1895, in the Senate from 1909 to 1915, and again in the House from 1921 to 1929. His quiet, but effective manner—"he worked more than he talked," wrote one admirer—and his thorough knowledge of finance, international relations, and water transportation won him the deep respect of his colleagues, but also hid his influential role in legislation from the eyes of historians. Author of several works on finance and corporations, as well as a biography of Senator John Sherman of Ohio, Burton served on the National Monetary Commission, and for two years was President of the Merchants National Bank of New York City. He made far fewer headlines than did the Roosevelt conservationists, but more than any other single man he was responsible for the failure of the multiple-purpose program.

Burton was frankly skeptical of the economic value of most of the new proposals, and he feared that logrolling would result in congressional approval of unsound projects. The Corps of Engineers agreed with him. To thwart the new ventures, Burton proposed that Congress create in the Corps of Engineers a special Board of Engineers for Rivers and Harbors which would exercise a final review over the feasibility of new projects.[8] This measure became

[7] Walter C. Cooper to Francis G. Newlands, Feb. 15, 1908 (FGN, Waterways-River Regulation, Correspondence, 1907–1911, 1).

[8] The editor of *Engineering News* hoped that this Board might eliminate logrolling and place the entire river and harbor program on a more systematic basis. *Engineering News* (April 17, 1902), 47, 310.

law in 1902. At the same time, although favoring the development of coastal rivers and harbors of easy access, Burton opposed inland waterway improvements and inland canal construction which, he argued, would not be worth the cost. Moreover, he committed the heresy of denying that waterway competition would effectively influence the level of railway rates. Until Burton retired as chairman of the Rivers and Harbors Committee he continued to exercise a restraining influence on the inland waterway movement.[9]

Waterways enthusiasts greeted this retrenchment policy with hostility. Unsuccessfully they had resisted the creation of the Board of Engineers on the ground that it would place in an executive agency power that should rest solely with Congress.[10] As the Corps turned in adverse reports on new projects, and as the Rivers and Harbors Committee turned down attempts to bypass the Corps, the bitterness increased and the new waterway enthusiasts resorted to more concerted political action and public appeals.[11] In 1901 the nation's waterway groups joined to form the National Rivers and Harbors Congress. Struggling on fitfully for a few years, this group reorganized in 1906 to become the most powerful national force for expanded waterway improvement.[12] It supported no specific project, but it crystallized the demands of many diverse groups by urging that Congress expand its annual river development appropriation to $50,000,000. At its December 1908 meeting the Rivers and Harbors Congress officially endorsed a $500,000,000 bond issue for river development, and demanded a waterway commission which would investigate and propose new projects.[13] To throttle the restraining influence of the Corps of Engineers once and for all, the Rivers and Harbors Congress president, Joseph E. Ransdell

[9] Burton's work is described in Pross, "Rivers and Harbors," 104–111.

[10] *New York Herald Tribune*, Jan. 26, 1901, cited in Pross, "Rivers and Harbors," 106.

[11] For example, see criticism of Burton in C. H. Claudy, "Circumventing Cape Hatteras; An Inland Waterway from Chesapeake Bay to Beaufort Inlet, North Carolina," *World To-day* (Dec. 1907), 13, 1248–1254.

[12] For a brief account of the 1907 Congress see *Forestry and Irrigation* (Jan. 1908), 14, 48–50.

[13] 60th Congress, 2nd Session, *Congressional Record*, 180.

of Louisiana, proposed to eliminate the Corps entirely by transferring its navigation work to a civilian department of public works.[14]

The Lakes-to-the-Gulf Deep Waterway

In this agitation the Lakes-to-the-Gulf Deep Waterway Association became especially prominent. It also became more closely connected with the Roosevelt water planners than did any other waterway group. Composed of manufacturers, wholesalers, and retailers throughout the Mississippi and Illinois river valleys, the Lakes-to-the-Gulf Association promoted a fourteen-foot channel from Chicago to New Orleans, sufficient to carry both lake- and ocean-going vessels.[15] This proposal expressed the aspirations of the entire Mississippi Valley, its hopes for increased trade and economic growth. As a result of Chicago's rapid development, its businessmen aspired to commercial leadership of the entire interior Valley.[16] On the other hand, St. Louis business leaders hoped to recapture trade lost when railroads had diverted the grain traffic to Memphis, a change which, they argued, had contributed largely to the decline of St. Louis as a river port.[17] Manufacturing concerns all through the Valley envisaged an ever-widening market to the south and southwest, rendered more accessible by lower transportation rates.[18] In conjunction with the Panama Canal, the Deep Waterway would open the entire west coast of South America to Midwest merchants.

[14] *Waterway Journal* (April 14, 1908), 21, 9.

[15] Harvey, "Lakes-to-the-Gulf."

[16] The Illinois Manufacturers' Association and the Chicago Commercial Association both led in advocating the project. See Chicago Commercial Association, *From the Great Lakes*; also statement of Lachlen Macleay, president of the Mississippi Valley Association, to the writer, Aug. 7, 1951.

[17] Inland Waterways Commission, *Minutes of the Meetings of the Inland Waterways Commission* (hereafter cited as Inland Waterways Commission, *Minutes*), May 17, 1907, testimony of J. F. Ellison (GP #2136).

[18] Chicago Commercial Association, *From the Great Lakes*, 17–19; WJ McGee to officers and members of the Latin-American Club and Foreign Trade Association, nd (probably April 1906) (WJM, Miscellaneous Letterbook #2).

In the lower Valley, consumers and cotton shippers were equally eager to share the benefits of low-cost river freight.[19]

Interest in the Deep Waterway arose originally from two centers, Chicago and St. Louis. Chicago and the state of Illinois became involved in the movement after the city in 1899 had completed construction of the Chicago Drainage Canal from Chicago to Lockport, Illinois, to divert Chicago sewage from Lake Michigan into the Illinois River. The state act of 1889 which authorized the canal specified that it be built for navigation as well. When it had completed this 24-foot waterway, the Chicago Sanitary District proposed that the state extend it from Lockport to the mouth of the Illinois River, ostensibly to aid navigation, but actually to solve a perplexing problem which the District faced. To divert water from Lake Michigan into the Illinois River would eventually produce eight times the normal flow of the river at its mouth, flood many acres of bottomland, and create damages for which the Sanitary District was legally responsible. The Deep Waterway, the District felt, would best solve this problem.[20]

Development of hydroelectric power was an integral part of this project. Through the sale of electricity the Sanitary District hoped to pay for the entire cost of its multiple-purpose project. By 1907 it had completed a generating plant of 43,000-horsepower capacity at the Lockport end of its canal. That same year it requested from the legislature permission to extend its channel five miles to the head of Lake Joliet and there develop 23,000 horsepower more. This phase of the Chicago project convinced many throughout the state that Illinois could develop the river for navigation and power, and pay for it with the revenue from the sale of electricity. In 1905 the General Assembly authorized an Internal Improvement Commission to investigate such a plan. Over 100,000 horsepower, the Com-

[19] Three of the five vice-presidents of the Lakes-to-the-Gulf Deep Waterway Association elected at its first meeting were from the lower Mississippi Valley —Arkansas, Mississippi, and Tennessee. Harvey, "Lakes-to-the-Gulf."

[20] *Engineering Record* (Mar. 24, 1900), 41, 265–266; Charles S. Deneen, "Vast Wealth for the State," *Technical World Magazine* (April 1908), 9, 127; Deneen, *Special Message to the Illinois Legislature* (Springfield, 1911), 12.

mission reported in 1907, lay undeveloped between Lockport and Utica. The state, it recommended, should retain ownership of these sites, lease them as a part of a combined power-navigation project, and use the proceeds to pay for the entire development. In November 1908 the voters of Illinois, in a popular referendum, approved a state bond issue to construct the canal as far as Utica. At the same time, Illinois groups petitioned Congress to complete the channel to St. Louis. Since they needed federal funds to finance their plan for state development, Illinois leaders became closely involved in the broader Lakes-to-the-Gulf agitation.[21]

Businessmen from St. Louis soon joined their Illinois neighbors. When the Lakes-to-the-Gulf Deep Waterway Association held its first convention in St. Louis, delegates from that city captured the organization and elected local men as its main officials.[22] The St. Louis Business Men's League spearheaded this drive, and within this organization James E. Smith, vice-president of the Simmons Hardware Company and resident consul in St. Louis for Japan, led the way. For many years Smith, as well as the entire League, had promoted the economic development of St. Louis. In 1903 they had sponsored the St. Louis Exposition; now they were ready for a more ambitious project.[23]

The Latin-American Club of St. Louis, composed of business leaders who eyed Latin-American markets, joined with the Business Men's League in promoting the Waterway. Anticipating the opening of the Panama Canal and the prospects of trading with the Orient as well as with South America, the Club in February 1907

[21] Deneen, *Special Message*, 6, 8; Deneen, "Vast Wealth for the State," 122; *Engineering Record* (April 7, 1900), 41, 313; Illinois Internal Improvement Commission, *Report* (Springfield, 1907).

[22] The organization's president was James Kerr Kavanaugh, of St. Louis, a coal-mine operator and a river traffic expert. The secretary was William F. Saunders, secretary of the Business Men's League.

[23] Walter B. Stevens, *History of St. Louis, the Fourth City* (St. Louis, 1909), 817; John W. Leonard, ed., *Book of St. Louisians* (St. Louis, 1906), 538. In the 1920's Smith was president of the Mississippi Valley Association, the organization which continued the agitation begun by the Lakes-to-the-Gulf Association.

sponsored a trip to Panama for twenty-four St. Louis businessmen to investigate future trade openings.[24] Secretary of State Elihu Root and John Barrett, the director of the International Bureau of American Republics, encouraged these new interests. Barrett, a frequent speaker at commercial conventions, was enthusiastic about the prospects. "The great Middle West," he declared, "will control the trade of the entire west coast of South America when the Panama Canal is completed, providing the deep waterway from the Lakes to the Gulf is constructed."[25]

In 1900 the Chicago Sanitary District petitioned Congress to construct the Deep Waterway, and two years later, in the Rivers and Harbors Act, Congress authorized a special Board of Army Engineers to survey the Illinois section. In their report in December 1905, the Engineers outlined a fourteen-foot channel to cost over $31,000,000. In the House, however, Representative Burton blocked the efforts of Valley congressmen to secure construction funds in the Rivers and Harbors Act of 1907.[26] At the same time, in January 1907, the Corps reported unfavorably on the fourteen-foot project below Cairo, Illinois. The cost, estimated at some $70,000,000, the Corps argued, would not be worth the commercial benefits. Instead, the report concluded, Congress should authorize an eight-foot waterway for the entire Illinois-Mississippi system.[27]

The entire middle Mississippi Valley turned its wrath upon Representative Burton and the Corps of Engineers. Since the distribution of money in the rivers and harbors bill was unfair, formally resolved the Missouri legislature, Burton was unfit to be chairman. The lawmakers of Illinois and Wisconsin acted in similar

[24] WJ McGee to officers and members of the Latin-American Club and Foreign Trade Association, nd (probably April 1906) (WJM, Miscellaneous Letterbook #2); *St. Louis Post-Dispatch*, Feb. 19, 1907 (GP #2137).

[25] William Flewellyn Saunders, "The President's Mississippi Journey," *American Review of Reviews* (Oct. 1907), XXXVI, 456–460.

[26] *Engineering Record* (March 24, 1900), 41, 265; 59th Congress, 1st Session, *House Document 263*; 59th Congress, 2nd Session, *Congressional Record*, 2028–2044, 2103–2117, 2301–2307.

[27] See comment on this report in *Engineering News* (Jan. 10, 1907), 57, 39.

fashion.[28] The St. Louis *Daily Globe-Democrat* attacked the appropriation of $19,000,000 in the Rivers and Harbors Act for improvement of navigation on the Great Lakes. This item, a fourfold increase, the editor argued, clearly indicated Burton's partiality for his own Cleveland congressional district.[29] The Corps of Engineers came in for equal condemnation. The Corps, many in the Valley complained, was unimaginative, unaware of the commercial possibilities of the Mississippi Valley, untrained in matters of trade and commerce. It approved projects in terms of economies in present traffic rather than faith in future commercial growth.[30] The *St. Louis Post-Dispatch* summed up the attack: "Hitherto the Atlantic and Pacific seaboards have controlled the policy of this country.... It is now up to the people of the Interior to demand their share of improvements."[31]

The Midwest's wrath because of the failure of its Deep Waterway project played an important role in its attack on the political power of the East in the Republican Party. Old-guard Republicans, Midwesterners thought, were the major bottleneck. Did not Republican congressional leaders carefully appoint to the Rivers and Harbors Committee men from the Great Lakes and seaport harbors, and exclude those from the upper Mississippi Valley? Between 1893 and 1915, for example, three Wisconsin representatives served a total of ten terms on the Committee. Each of these, however, came from a district bordering on Lake Michigan and none came from one on the Mississippi. Not until 1915 was a Mississippi Valley congressman from Wisconsin, James Frear from the Tenth District, appointed to the Committee.[32] Reflecting this membership, the

[28] *St. Louis Republic*, Jan. 31, 1907 (GP #2136).

[29] *St. Louis Daily Globe-Democrat*, Feb. 20, 1907 (GP #2137).

[30] *Proceedings, Lakes-to-the-Gulf Deep Waterway Association*, 1910, 37–39; John L. Mathews, "The Future of Our Navigable Waters," *Atlantic Monthly* (Dec. 1907), 100, 723–724.

[31] *St. Louis Post-Dispatch*, Apr. 7, 1907 (GP #2137).

[32] Between 1893 and 1919, of 248 Congress-units of membership on the Rivers and Harbors Committee, 161 or 65 per cent came from districts bordering on the seacoast or the Great Lakes. 24 per cent came from the upper Ohio and Mississippi flood plain areas. Only 8·5 per cent came from

Rivers and Harbors Committee supported funds to improve Eastern seaports and the Great Lakes ore routes, but refused to undertake the newer projects for developing inland waterways.[33] Despite the fact that the leading apostle of retrenchment, Speaker of the House "Uncle Joe" Cannon, came from Illinois, he became a major target of attack for the Deep Waterway enthusiasts.

The Multiple-Purpose Movement

This enthusiasm for waterway development dovetailed nicely with the expanding views of hydrologists and engineers in the Roosevelt administration. Earlier experience with reclamation and forestry affairs seemed to point toward a multiple-purpose river basin program. The Geological Survey had brought forward the concept of water as a single resource with many uses. The Reclamation Service was constructing reservoirs to control water for irrigation, and had become aware of the possibilities of combining irrigation storage with hydroelectric power production. The Reclamation Service developed power at its first reservoir project, and Congress in 1906 authorized that agency to undertake its general development and sale.[34] Practical experience with these problems prompted officials of the Roosevelt administration almost inevitably to think in larger terms.

The enormous possibilities of basin-wide river development suddenly captured the imagination of Newell, Pinchot, Garfield, and other conservation leaders. Flood waters, now wasted, could, if harnessed, aid navigation, produce electric energy, and provide water for irrigation and industrial use. It also became clear to these men that maximum development required multiple-purpose development. Engineering works which tapped a river for one use

---

the upper Mississippi, the Missouri, the Illinois, and the middle Mississippi, and, widely scattered in time, they were easily outnumbered.

[33] See speech of Joseph G. Cannon at the Fifth Annual Meeting of the National Rivers and Harbors Congress, Dec. 1909, as quoted in *Conservation* (Jan. 1909), 15, 51–52.

[34] For hydroelectric development at the Salt River, Arizona project, the first power plant to be installed, see James, *Reclaiming the Arid West*, 65–85.

alone might rule out other uses which could yield even greater benefits. A low dam for navigation, for example, might prevent construction of a higher dam at the same site that would produce hydroelectric power as well. Senator John H. Bankhead, of Alabama, who worked closely with the administration on this problem, described the intimate link between these two uses:

> The Government of the United States, desiring to improve the navigability of a river where great and valuable water power may be established, ought in the very inauguration of the work to look to it that in the construction and operation of locks, power at the same time will be provided. This they must do . . . unless the Government makes up its mind that for all time this power, valuable as it may be, must give place to navigation exclusively.[35]

The multiple-purpose concept required attention to the entire river basin as well as to the size and design of reservoirs. Failure to control erosion and silt might impair the long-range value of engineering works.[36] The multiple-purpose approach, therefore, brought together federal officials in both land and water agencies in a common venture.

Past experience with conservation problems convinced the administration that in river development the federal government should take the lead. Only Congress could provide the funds; inadequate state and private finance had forced the federal government to undertake irrigation, and an expanded water program would cost far more. The interstate character of streams also required federal jurisdiction for development, which should follow "the limits imposed by nature rather than the artificial lines defining the sovereignty of the various states."[37] Reclamation Service plans to control rivers in the West had gone awry when upstream communities resisted water storage in their states for downstream use.[38] Only the federal government could overcome these interstate bickerings

[35] 60th Congress, 2nd Session, *Congressional Record*, 3358.

[36] WJ McGee, "Our Great River," *The World's Work* (Feb. 1907), 13, 8576–8584.

[37] *Ann. Rept. USGS, 1907*, 7.

[38] Roosevelt to William E. Curtis, May 19, 1906 (TR).

and bring about the fullest development of interstate streams. Since the federal government had jurisdiction over the public lands and navigable streams, only federal law could prevent speculation in crucial sites from tying up development.[39] Flood control, clearly an interstate problem, required interstate treatment. If Congress could be persuaded to overcome its legalistic attachment to states' rights and grant considerable authority to an administrative agency, the federal government could promote untold economic growth that would benefit every local community in the nation.

Agitation for inland waterways provided the opportunity for federal leaders to carry out their multiple-purpose views. Waterway associations supported only their particular navigation projects; they displayed little interest in larger ideas. But administration leaders hoped that they could marshal this interest in limited plans to support their broader objectives, and for a time it appeared that they might succeed. WJ McGee linked these two groups. Geologist, anthropologist, philosopher, former member of the Geological Survey, and assistant to John Wesley Powell, the head of the Bureau of Ethnology, McGee became the chief theorist of the conservation movement.[40] He also became one of its most crucial promoters within the Roosevelt administration and in the country at large. Almost daily he presented new ideas to Roosevelt, Pinchot, and Garfield; he drew up presidential letters and messages, formulated policies, and organized conferences. He spoke frequently at waterway meetings throughout the nation, channeled their efforts into the administration's program, and planted those few seeds of multiple-purpose planning which took root in the minds of waterway enthusiasts.

A native of Dubuque County, Iowa, born in 1853, McGee was a self-made scientist.[41] While working on a farm and at the forge in

[39] Frederick H. Newell, "Memorandum concerning water power presented to the Inland Waterways Commission, April 1908" (GP #2135).

[40] See McGee's views in "Water as a Resource," in "Conservation of Natural Resources," *Annals of the American Academy of Political and Social Science* (May 1909), 33, 37–50; McGee, "Our Great River."

[41] For a brief account of McGee's life see Washington (D.C.) Academy of Sciences, *The McGee Memorial Meeting of the Washington Academy of Sciences* (Baltimore, 1916). A highly laudatory account of McGee's influence

his teens and twenties, he read widely in mathematics and natural science. As a private citizen between 1877 and 1881 he made an extensive geologic and topographic survey of northeastern Iowa. Coming to Washington in 1885, he soon became a leader in scientific circles in the nation's capital. He assisted in organizing both the National Geographical Society and the Geological Society of America, and for many years edited the journals of those two organizations. He was president of the National Geographical Society in 1904–05 and of the American Anthropological Society in 1911. Always an intense optimist, McGee was most vitally interested in applied science. Though a theorist of considerable force, he took up most enthusiastically those ventures which could yield immediate material benefit to man. In this light he viewed the entire conservation movement. When McGee died of cancer in September 1912, Washington scientists honored him with an elaborate memorial meeting.

McGee first came into contact with the waterway movement in St. Louis. In 1903 he had left the Bureau of Ethnology to become chief of the anthropological section of the St. Louis Exposition. Here he brought together people, commercial products, and archeological remains from every corner of the globe. Here also he came to know the city's commercial and business community which sponsored the event. After the Exposition McGee took charge of the new St. Louis Museum and inaugurated a program to establish there a center for economic, anthropological, and archeological displays from the entire Southwest and Latin America.[42] "Every consideration of geography and ethnology no less than of immediate commerce," he declared to the city's businessmen, "commands our citizens to extend their enterprises beyond the Rio Grande and the thirtieth parallel, even beyond the Gulf and the Caribbean, to afar across the equators. . . . Saint Louis is the natural key city not only to our own

---

on the conservation movement is in Pinchot, *Breaking New Ground*, 358–360. See also Whitney R. Cross, "WJ McGee and the Idea of Conservation," *Historian* XV (Spring 1953), 148–162.

[42] McGee to Mrs. Aaron Morley Wilcox, Sept. 28, 1905 (WJM, General Letterbook #1).

Southwest but to all Latin-America and much of the Pacific. . . ."[43] In April 1906 the Latin-American Club, for which McGee had become a chief spokesman, sent him as its representative to a convention of Southwestern commercial organizations, and in October to the first meeting of the Lakes-to-the-Gulf Deep Waterway Association.[44]

McGee's fertile and imaginative mind went far beyond the commercial possibilities of the Mississippi River and its trade with the Southwest. In fact, he became discouraged with the limited views of the Deep Waterway Association and its failure to take up the multiple-purpose idea. The proceedings at the Convention, he lamented to Representative Joseph E. Ransdell, dealt almost wholly with navigation, "the aspect in which the waterway is merely a body of liquid as is a lake or sea"; the delegates did not discuss "either . . . the engineering aspect of the project [or] . . . the geologic aspect—i.e., . . . the river considered as an agency engaged in the transportation of silts, sands, and other solid matters."[45] In February 1907 he outlined his broader views in an article in *The World's Work*. Each stream, he wrote, is "an interrelated system in which the several parts are so closely interdependent that no section can be brought under control without at least partial control of all other portions. . . . It is in this concept of the river as a power to be controlled by engineering projects, and at the same time as an agency of interdependent parts, that the views of the engineer and the geologist must meet and merge. . . ."[46] McGee had described with great accuracy the very process whereby the views of resource leaders in the Roosevelt administration had met and merged.

McGee worked quickly and energetically to implement his views.

[43] McGee to the officers and members of the Latin-American Club and Foreign Trade Association, nd (probably April 1906) (WJM, Miscellaneous Letterbook #2); press release by McGee, c. Oct. 9, 1905 (WJM, General Letterbook #1).

[44] McGee to James M. Arbuckle, Oct. 22, 1906 (WJM, Miscellaneous Letterbook #3).

[45] McGee to Joseph E. Ransdell, Nov. 24, 1906 (WJM, Miscellaneous Letterbook #3).

[46] McGee, "Our Great River."

He prevailed upon the president of the National Rivers and Harbors Congress, Joseph E. Ransdell, to invite Frederick H. Newell of the Reclamation Service to speak at the group's annual convention in December 1906. The work of the Reclamation Service, McGee felt, most clearly exemplified multiple-purpose river control. McGee, in turn, urged Newell to speak on the effect of irrigation works on the flow of rivers downstream.[47] During his trip to the Rivers and Harbors convention, McGee discussed with President William K. Kavanaugh and other leaders of the Deep Waterway Association a plan whereby President Roosevelt would appoint a commission to explore the possibilities of river development.[48] Kavanaugh quickly grasped the implications of McGee's ideas. In January, when the Corps of Engineers reported adversely on the Mississippi project, officials of the Deep Waterway Association, prompted by the St. Louis Business Men's League, took up McGee's plan for a waterway commission as the best hope of bypassing the hostile Rivers and Harbors Committee and the Corps of Engineers. In December, McGee had conferred with several administration leaders in Washington; by January, he had discussed his views with Gifford Pinchot. On February 10 he left St. Louis again with petitions asking the President to create an Inland Waterways Commission. On March 12 he formally presented these to Roosevelt, and two days later the President announced the appointment of the Commission.

The Inland Waterways Commission

President Roosevelt, in appointing the Inland Waterways Commission, did not follow McGee's plan literally. McGee had selected a commission of four experts, one from each of the federal departments involved in water problems: Gifford Pinchot from the Forest Service, Frederick H. Newell from the Reclamation Service, Brigadier-General Alexander Mackenzie, Chief of the Corps of Engineers, and Lawrence O. Murray, Assistant Secretary of the

[47] McGee to Ransdell, Nov. 24, 1906 (WJM, Miscellaneous Letterbook #3); McGee to Newell, Nov. 19, 1906 (WJM, Miscellaneous Letterbook #3).

[48] McGee to W. K. Kavanaugh, Apr. 3, 1907 (WJM, Miscellaneous Letterbook #3).

Department of Commerce and Labor. The President, more aware of the Commission's political ramifications, added to this group two members of the House, Theodore E. Burton of Ohio and John H. Bankhead of Alabama, and two from the Senate, Francis G. Newlands from Nevada and William Warner from Kansas. He replaced Murray with Herbert Knox Smith, Commissioner of Corporations, and, after appointing McGee to a post in the Bureau of Soils so that he would be eligible, selected him as the Commission's Secretary.[49]

The climate of opinion which surrounded the Commission from the day of its appointment differed radically from any previous approach to water problems. Embodying the ideas of a first draft which McGee had written, Roosevelt made clear in his letter of appointment that he intended the Commission to take a multiple-purpose viewpoint. He wrote:

> Works designed to control our waterways have thus far usually been undertaken for a single purpose, such as the improvement of navigation, the development of power, the irrigation of arid lands, the protection of lowlands from floods, or to supply water for domestic and manufacturing purposes. . . . The time has come for merging local projects and uses of the inland waters in a comprehensive plan designed for the benefit of the entire country. Such a plan should consider and include all the uses to which streams may be put, and should bring together and co-ordinate the points of view of all users of water.[50]

From this point on, the President threw his prestige behind the movement for multiple-purpose river development; it became identified as one of the "Roosevelt policies."

Beginning its work in April 1907, the Inland Waterways Commission devoted most of its active investigation to the problem of river navigation, why it had declined, and how it could be revived. In May the Commission traveled by steamboat down the Mississippi, stopping at St. Louis, Memphis, and other river ports to question

[49] See proposed letter of appointment, Feb. 4, 1907 (WJM, Miscellaneous Letterbook #3) and Roosevelt to members of the Inland Waterways Commission, March 14, 1907 (TR).

[50] *Ibid.*

veterans of river commerce and officers of the Corps of Engineers. In September it began a tour of the Great Lakes, stopped at St. Paul, and joined President Roosevelt at Keokuk, Iowa for his widely publicized jaunt down the River. Some members of the Commission personally investigated the Columbia River; others explored the problems of the Central Valley of California. The Commission appointed a sub-committee to consider water power, examined several proposals to control Mississippi River floods, and inquired into the relation of forest cover to stream flow. On February 3 it signed its report, and on February 28 the President forwarded the document to Congress.[51]

The work of the Inland Waterways Commission brought to the fore a figure new to the conservation scene—Marshall O. Leighton, the Chief Hydrographer of the Geological Survey. While McGee formulated broad ideas of multiple-purpose development, Leighton devised the specific engineering plans to make the scheme practical. Historians have forgotten Leighton even more than they have overlooked McGee, for in retrospect the Roosevelt conservation leaders gave the latter due credit but failed even to mention the work of the Chief Hydrographer. Yet it was Leighton who devised the first concrete plan for multiple-purpose river development, long before the Tennessee Valley Authority emerged as an idea. He transformed McGee's vision into a practical blueprint.

Born in Maine in 1874, Marshall O. Leighton graduated from the Massachusetts Institute of Technology in 1896. Six years later he entered the U. S. Geological Survey as a resident hydrographer. Rising quickly in the service, he became Chief Hydrographer in 1906, which position he held until 1913 when he resigned to become a private consulting engineer. Between 1903 and 1906 Leighton served on two northern New Jersey Flood Commissions which tackled the problem of Passaic River floods. This experience convinced him that the same reservoirs that produced hydroelectric power could control floods, and turned him toward the views of

[51] A full account of the Commission's activities is contained in the minutes of its meetings in GP#2136.

Newell and McGee.[52] Leighton became advisory hydrographer to the Inland Waterways Commission, and, at its request, drew up a multiple-purpose plan for the Ohio River system, consisting of one hundred reservoirs to impound the flood waters of the entire basin.[53] By producing and selling the power from these reservoirs, Leighton argued, the federal government could finance the entire scheme.[54]

The Inland Waterways Commission approved his report, and emphatically supported the multiple-purpose idea. "Hereafter," it recommended, "plans for the improvement of navigation in inland waterways . . . should take account of the purification of the waters, the development of power, the control of floods, the reclamation of lands by irrigation and drainage, and all other uses of the waters or benefits to be derived from their control."[55] This proposal, the Commission continued, required a single executive agency to coordinate all water resource administration. Roosevelt especially emphasized this point in his message accompanying the report: "No single agency has been responsible under the Congress for making the best use of our rivers, or for exercising foresight in their development. . . . We shall not succeed until the responsibility for administering the policy and executing and extending the plan is definitely laid on one man or group of men who can be held accountable."[56]

The U. S. Army Corps of Engineers was the special target of these views. Its engineers disagreed violently with the new turn of affairs in water development. They refused to consider seriously a multiple-purpose approach.[57] They attacked the theory that forests

[52] Leighton statement to Samuel P. Hays, Sept. 6, 1951.

[53] Inland Waterways Commission, *Minutes*, Sept. 25, 1907. These events as narrated by Leighton are included in his manuscript, "Conservation and Water Resources," in GP #3200.

[54] 60th Congress, 1st Session, *Sen. Doc. 325, Preliminary Report of the Inland Waterways Commission* (hereafter cited as IWC, *Report*), 451–490.

[55] IWC, *Report*, 19.

[56] *Ibid.*, 5.

[57] As an example, see E. W. Burr to the Secretary of War, June 22, 1910, NA, War, RG #77. See also Philip P. Wells, "Water Power—Regulation and Administration" (GP #1665); *Report of Commission of Public Works to the Governor of California, 1895*, 138, referred to by Carl E. Grunsky, in Grunsky, Chittenden, and Labelle, "The Flood of March, 1907."

retarded run-off, or that engineers could devise an economical reservoir system.[58] The Roosevelt administration had formulated its new approach to water development without the advice of the Corps, and in spite of its quiet opposition. The Engineers well knew that a new water commission with power to act would destroy their autonomy and subordinate them to men with an entirely different viewpoint. The Chief of the Corps, Brigadier-General Alexander Mackenzie, alone dissented from the report of the Inland Waterways Commission. River planning, he argued, should subordinate flood control, hydroelectric power, and irrigation to navigation. He denied that federal river programs required greater coordination. "All necessary cooperation," he concluded, "can be equally well provided for by the existing agencies of the Government. . . ."[59]

Senator Newlands Presents a New Bill

In December 1907 Senator Francis G. Newlands presented to Congress a bill to carry out the recommendations of the Inland Waterways Commission. A permanent body appointed by the President would continue to investigate water problems, authorize projects, supervise construction, and coordinate the activities of all federal water resource agencies. To finance the work Congress would establish an Inland Waterway Fund of $50,000,000, and the President could restore it to that amount whenever it fell below $20,000,000. The Commission could draw upon this Fund at its own discretion, without annual authorization from Congress.[60]

The Newlands measure anticipated a far larger grant of power to an administrative agency than Congress had heretofore authorized for water development, save for the Reclamation Act of 1902.[61] To Newlands, however, the success of the new waterway program

[58] For example, see Hiram M. Chittenden, "Forests and Reservoirs in their Relation to Stream Flow with Particular Reference to Navigable Rivers," *Proceedings, American Society of Civil Engineers* (Sept. 1908), 34, 924–997.

[59] IWC, *Report*, 30–31.

[60] 60th Congress, 1st Session, *Congressional Record*, 389–400.

[61] For Newlands' views on the Reclamation Act of 1902 and its larger implications see Newlands to Dent H. Robert, Sept. 17, 1904 (FGN, Letters).

depended entirely upon the Commission's power to select projects and to draw upon a source of funds independent of annual congressional appropriation, precisely as the Reclamation Act had provided. To his reluctant colleagues in the Senate he argued:

Large powers and a comparatively free hand should be given to an administrative body of experts in the full development of projects, lest the complexity of the transaction, the time necessary to secure Congressional approval, the difference in view as to purpose or method, may result in indecision and delay, the worst enemies of effective development. . . . I hope to see a commission of experts that will have the power to initiate both investigation and construction and with ample funds to complete its projects, not a commission that will have to wait on the tardy initiative of Congress as to projects the details of which it is incapable of dealing with. . . . I wish it to handle this capital just as a board of directors of a great corporation would do. . . .[62]

The Roosevelt administration readily agreed with these views. Congress, far more responsive to the demands of local constituents than to the requirements of scientific planning, could not select projects rationally. It preferred to carry forward many projects at once, though with meager funds for each, rather than take them up in the priority of their greatest economic value.[63] Unless a multiple-purpose program could circumvent these practices, the water conservationists argued, it would fail.

The Newlands bill met with a cool reception in the Senate. Under prodding from Newlands, the Committee on Commerce, to which the bill was referred, finally appointed a special sub-committee to consider it. This group agreed on the vital point of granting the Commission a comparatively free hand within the limits of the fund.[64] Secretary of War Taft, in charge of the Corps of Engineers,

[62] IWC, *Report*, 31; 60th Congress, 1st Session, *Congressional Record*, 391.

[63] Testimony of Henry L. Stimson before House Select Committee on the Budget, 66th Congress, 1st Session, *Hearings on the Establishment of a National Budget System*, 641.

[64] Newlands to Burton, Feb. 29, 1908 (FGN, Waterways-River Regulation, Correspondence, 1907–1911, 2).

and the Inland Waterways Commission sent in favorable reports.[65] The sub-committee, in turn, reported the bill favorably to the entire Committee, but there it died.

The favorable messages from Taft and the Inland Waterways Commission did not reveal the true attitude of either. In fact, their failure to support the measure fully played an important part in its defeat. Taft sent in a favorable report only under pressure from the President. Roosevelt knew that the Commerce Committee would refer the bill to the War Department, that in the normal routine it would go to General Mackenzie, Chief of the Corps, who had dissented from the conclusions of the Inland Waterways Commission, and that Taft, agreeing with the Corps' attitude toward water development, would accept the Chief's recommendations. To counteract the influence of the Corps and obtain a favorable report on the bill, Roosevelt wrote to the War Department on February 29, 1908:

> Senator Newlands' bill in reference to the Waterways Commission will doubtless be sent to the War Department for a report. In the regular routine this would go to General Mackenzie. This is undesirable. The bill should not go to any of the engineers. It represents in a general sense the policy of the Commission, which policy is mine, but which policy is not the one approved by General Mackenzie and the engineers. Therefore I desire that the bill be sent direct to me, and I will, after a conference with the Secretary of War, direct what answer shall be sent in reference thereto.[66]

The Secretary of War complied with the President's wishes. Although he revealed between the lines of judicial reasoning his misgivings with the Newlands program, he permitted Pinchot and the Forest Service law officer Philip Wells to criticize his draft, and sent in a favorable message.[67]

[65] Mimeographed copy of report of Secretary of War on S 500 in GP #1665; Burton to Committee on Commerce, Apr. 20, 1908 (GP #1666).

[66] Roosevelt to the War Department, Feb. 29, 1908 (TR).

[67] See copy of Taft report on S 500, in draft form, showing many penciled changes in Pinchot's and Wells' handwriting, in RG #95, Law Office, Correspondence files, Drawer 36, folder "Muscle Shoals."

The Corps was furious that the President had used Taft's prestige to support a policy they disapproved. Through Mackenzie's dissent to the recommendations of the Inland Waterways Commission, it thwarted Roosevelt's efforts and presented its hostile views to the Committee on Commerce.[68] Members of the Corps tried to prevent circulation of the Taft report. In January 1909, for example, the new Chief of the Corps, Brigadier-General William L. Marshall, replied to a request for a copy of the message: "The report mentioned is not on the files of this office, nor is there any record that such a report was made by the Secretary of War."[69] Conservationists, however, eagerly publicized Taft's statement. The Lakes-to-the-Gulf Deep Waterway Association used it to gain prestige for the Newlands measure.[70] McGee urged the Nevada Senator to emphasize that the views of the Corps were "at disagreement not only with the report and policy of the Commission but with the report and policy of the Secretary of War and with the views of the Executive."[71] In reply, Newlands inserted a copy of the Taft message in a public document describing the Newlands measure.

Roosevelt hardly succeeded in nullifying the Corps' adverse influence. The Inland Waterways Commission, on the other hand, could not hide the opposition of its Chairman, Representative Theodore E. Burton of Ohio. Burton refused to grant to the proposed water board the power either to authorize public works or to provide funds for their construction. In this he spoke not only for himself, but for the prevailing sentiment in Congress. The lawmakers as a whole opposed an independent body that would take

[68] The day after Burton reported to the Committee on Commerce the views of the Inland Waterways Commission on the Newlands bill (Apr. 20, 1908), he again wrote the Committee to point out specifically that Mackenzie had dissented from the views of the majority of the Commission. Burton to Committee on Commerce, Apr. 21, 1908 (GP #1666).

[69] Brigadier-General William L. Marshall to Secretary of War, Jan. 23, 1909 (RG #77, #70392).

[70] W. F. Saunders to McGee, Apr. 27, 1908 (FGN, Waterways-River Regulation, Correspondence, 1907–1911, 2).

[71] McGee to Newlands, Apr. 21, 1908 (FGN, Waterways-River Regulation, Correspondence, 1907–1911, 2).

from Congress the power to decide each project, while the Republican leaders were appalled by the large increase in federal expenditures which the Newlands measure contemplated. Burton's strategic position as Chairman of the Rivers and Harbors Committee and his general influence in the House as an authority on water problems persuaded the Inland Waterways Commission to follow his views. Its report clearly approved the principle of coordination in water resource administration, but, it went on, "the specific mode of providing means for improving and promoting navigation should be left to the wisdom of Congress."[72]

During the early months of 1908 Newlands and Burton sparred over the precise wording of the measure. "The success of this work," the Senator wrote, "will depend upon continuous and consecutive work, with ample funds, through an administrative body with a comparatively free hand." The bill should read that the Commission could undertake any project "when approved by the President." Burton, in reply, offered a substitute, that the Commission could act only "as authorized by Congress."[73] Knowing full well Burton's power in the House, the Senator soon capitulated. He persuaded the Senate sub-committee to modify the measure so as to require congressional authorization before the proposed Commission could undertake any project it had found feasible.[74]

Representative Burton refused to support even this compromise version. Instead he brought forward a measure to authorize the President to continue the Inland Waterways Commission until July 1, 1909 and require it to report again in December 1908.[75] By this stroke Burton held out a ray of hope to the Newlands contingent but staved off a congressional commitment to any definite policy. The President and his advisers fumed at this rough treatment of

[72] Burton to Committee on Commerce, U. S. Senate, Apr. 20, 1908 (GP #1666).

[73] Newlands to Burton, Feb. 29, 1908 (FGN, Waterways-River Regulation, Correspondence, 1907–1911, 2).

[74] Newlands to Burton, Apr. 30, 1908 (FGN, Waterways-River Regulation, Correspondence, 1907–1911, 2).

[75] 60th Congress, 1st Session, *Congressional Record*, 6419–6420.

the Newlands bills, but they faced the political realities and backed the Burton substitute. Roosevelt tried to persuade Cannon to support it. "Of course," he threatened, "in any event I would continue the present Commission, but for me to do so when Congress had failed to recognize it would serve to emphasize an omission by Congress as to a matter in which I think the country is very deeply interested."[76] When it appeared that Congress would limit the personnel of the Commission to congressmen alone, the administration lost all interest and hoped that the bill would fail.[77] It did pass the House on May 16, 1908 by a vote of 226–2, but died in the Senate.[78]

A New Water Power Policy

Hydroelectric power provided the financial key to the entire multiple-purpose plan. Through sale of power, as Marshall O. Leighton outlined to the Inland Waterways Commission, the federal government could pay for river development without new appropriations from Congress. The Commission approved this principle, but its members at first differed on the method of applying it.[79] Some toyed with a proposal that private companies contribute that part of the cost of the multiple-purpose dam allocated to power, and the public agencies that part allocated to navigation and flood control.[80] When the Corps of Engineers and many in Congress objected that this would give rise to conflicts in operation and administration, the Commission agreed that the federal government should contribute the entire capital cost and lease power rights to private parties. As early as 1903, in a veto of the Muscle Shoals bill, President Roosevelt pointed out the possibilities of using power revenue to finance navigation improvement. "It seems clear," he wrote at that time, "that

[76] Roosevelt to Cannon, May 13, 1908 (TR).

[77] Pinchot to Elbert F. Baldwin, May 27, 1908 (GP #2137).

[78] *Ibid.*, 6420–6424. See also Roosevelt to Newlands, May 18, 1908 and Roosevelt to Burton, May 18, 1908 (both in TR); Newlands to Roosevelt, May 20, 1908 (FGN, Waterways-River Regulation, Correspondence, 1907–1911, 2).

[79] IWC, *Minutes*, Apr. 16, 1908 (GP #2136).

[80] 60th Congress, 2nd Session, *Congressional Record*, 3358.

justice to the taxpayers of the country demands that when the Government is or may be called upon to improve a stream, the improvement should be made to pay for itself, so far as practicable."[81] Five years later the administration worked out a plan for federal development and lease of water power on all navigable streams.

This new water power policy evolved within a maze of existing legislation; in fact, for its legal support the administration relied upon new interpretation of measures already on the statute books. In 1890 and 1899 Congress passed laws regulating the erection of dams in navigable streams to prevent them from interfering with river shipping.[82] The 1899 statute required that Congress (or a state if the stream lay within its borders) approve every such dam, and that the Chief of the Corps of Engineers in both cases approve detailed plans and specifications. During the early years of the twentieth century, the number of special acts for power licenses increased rapidly.[83] To cut down on its work Congress passed the General Dam Act of 1906 establishing in detail the conditions under which dams could be erected, but requiring that Congress approve each specific grant.[84]

Under this 1906 law Roosevelt signed twenty-five special acts permitting private parties to erect dams in navigable streams.[85] In the winter of 1907–08, however, the President began to feel that he should sign no more such measures unless they provided for some financial return to the government, regulation of rates, and recapture of the dam in case the corporation violated the terms of the grant. The earlier bills he had signed had contained no such stipulations. The precise reasons for the President's change of view remain obscure; however, it seems logical that Pinchot, who during

[81] 57th Congress, 2nd Session, *Congressional Record*, 3071.

[82] A good review of water power legislation is in Kerwin, *Water-Power Legislation*, 108.

[83] The number of special water-power acts by years was: 1903 (1); 1904 (6); 1905 (4); 1906 (17). These figures are compiled from Kerwin, *Water-Power Legislation*, Appendix IV, 333–340.

[84] 58th Congress, 3rd Session, *Congressional Record*, 3886.

[85] Kerwin, *Water-Power Legislation*, 114–115.

1907 worked out his permit system for the national forests, probably influenced him. In the spring of 1908 the Inland Waterways Commission appointed a committee of three—Newell, Pinchot, and McGee—to consider water power development on navigable streams.[86] A short time later the President moved to assert his new policy. In a special message to Congress on February 26, 1908 he attacked bills then pending which provided for perpetual grants without rentals or their equivalent.[87] On March 16, 1908 Roosevelt informed the Secretary of War and the chairmen of the respective Senate and House Committees that he would "sign no more bills hereafter which do not provide specifically for a charge and a definite limitation in time of the rights conferred."[88] His first veto came on April 13, 1908. Finding that this case presented special circumstances, however, he later recommended that Congress pass the bill over his veto. His second veto, on January 15, 1909, he did not change.[89]

The President had to reckon with advisers other than Pinchot, Newell, and McGee. Secretary of War Taft and the Corps of Engineers, in charge of administering laws affecting navigable streams, disagreed with the new power policy. The two groups clashed over the interpretation of the General Dam Act of 1906. According to

[86] IWC, *Minutes*, Feb. 3, 1908 (GP #2136). Commission proceedings concerning water power can be traced through the *Minutes*.

[87] 60th Congress, 1st Session, *Sen. Doc. 325*, v. The precise period during which Roosevelt changed his views on this question is not yet known. As late as March 4, 1907 Roosevelt had signed a bill assigning to the Alabama Power Company a perpetual grant for water power development in connection with a lock on the Coosa River in Alabama. The President accepted Pinchot's position on water power at least with respect to the public lands prior to February 1, 1908. See Richard M. Saltonstall to Pinchot, Feb. 1, 1908 (GP #1666).

[88] Roosevelt to Secretary of War, Mar. 16, 1908; Roosevelt to William P. Frye, Mar. 18, 1908; Roosevelt to Theodore E. Burton, Mar. 18, 1908 (TR).

[89] 60th Congress, 1st Session, *Congressional Record*, 4698–4699; Roosevelt to J. Adam Bede, May 7, 1908 (TR); James R. Garfield to Frederick V. Stevens, Committee on Interstate and Foreign Commerce (House), May 23, 1908 in memorandum prepared by Roosevelt, May 26, 1908 (TR); 60th Congress, 2nd Session, *House Doc. 1350*.

that Act, in approving plans and specifications, the Chief of Engineers and the Secretary of War could "impose any condition and stipulation that they deem necessary to protect the present and future interests of the United States."[90] This clause, Pinchot, Newell, and McGee contended, conveyed as broad a grant of power to an administrative agency as had the Right-of-Way Act of 1901. Roosevelt agreed. The General Dam Act, he wrote the Secretary of War in March 1908, "would seem to authorize the Government to make a reasonable charge for the valuable rights it grants."[91] Lawyers from the War Department strenuously objected. The Constitution, they argued, granted power to Congress to regulate dams only for navigation, and both Taft and the Corps shared this view.[92]

The President could not persuade Taft or the Corps to change their opinion, but he did take action to prevent them from shaping administrative policy; he instructed the Secretary and the Corps to accept the views of Pinchot, McGee, and Newell as the official position of the War Department, and took steps to see that they would comply. For example, late in March the President ordered Taft to obtain the advice of Gifford Pinchot before he rendered an opinion on water power bills which Congress referred to the War Department for its comments. Roosevelt indicated to Taft that Pinchot's opinion should become the War Department's opinion.[93] Pinchot's law officer, Philip P. Wells, in turn, prepared a draft clause providing for a limited grant and water power charge, and on April 10 forwarded it to Brigadier-General Alexander Mackenzie, Chief of the Corps. Pinchot, Wells informed the Chief, desired that the Corps automatically insert this clause in each new water power bill.[94] From this time on, bills returned to Congress from the War

[90] Act of June 21, 1906 (34 *Stat.* 386).

[91] Roosevelt to Secretary of War, Mar. 16, 1908 (TR).

[92] IWC, *Minutes*, Sept. 26, 1907 (GP #2136).

[93] William H. Taft, "Memorandum for the Assistant Secretary," Apr. 1, 1908 (WHT, Secretary of War, Letterpress Book #65); John C. Scofield, Chief Clerk of the Corps of Engineers to Fred W. Carpenter, secretary to President Taft, July 13, 1909 (RG #77, #49678/26).

[94] Wells to Mackenzie, Apr. 10, 1908 (RG #95, Law Office, Correspondence Files).

Department reflected accurately the administration's new policy.[95]

At the same time the President took steps to make sure that specifications which the Corps, under the General Dam Act, approved for each new installation, would also provide for charges and a limited grant. On June 15, 1908 Roosevelt ordered the Secretary of War to submit each new plan for hydroelectric dams to the Attorney-General; he ordered the Attorney-General to submit these plans, in turn, to George Woodruff, Assistant Attorney-General for the Department of the Interior.[96] Because of Woodruff's role in developing water power policy for the public lands, Roosevelt could rely with confidence on his judgment. In spite of these elaborate arrangements, the Corps forwarded to Woodruff no plans for new dams. In September, he became suspicious that the Corps had not complied with the President's request; but when he inquired if the Corps had understood the new procedure correctly, the Chief Engineer replied that he had approved no new dams since the order had gone into effect.[97] There is no evidence that the Corps actually changed any plans and specifications as a result of advice rendered by Assistant Attorney-General Woodruff.

These administrative devices were, to say the least, unorthodox. They forced the Secretary of War to rely on other departments for his official attitude on matters within the jurisdiction of the War Department. The President even communicated with congressional committees about water power in navigable streams, not through the Secretary of War but through Secretary of the Interior Garfield.[98]

[95] This entire procedure can be followed in a series of letters and indentures in the records of the Corps of Engineers and the War Department from June 16, 1908 to April 25, 1910 (RG #107, #15509).

[96] Roosevelt to Secretary of War, June 15, 1908; Roosevelt to Attorney-General, June 15, 1908 (both in TR).

[97] Woodruff to William Loeb, Jr., Sept. 29, 1908; Loeb to Luke E. Wright, Sept. 29, 1908; Loeb to Woodruff, Sept. 29, 1908; Robert Shaw Oliver to Loeb, Oct. 7, 1908; Loeb to Woodruff, Oct. 8, 1908 (all in RG #48, file 1–70, Conservation of Natural Resources, "Dam Privileges").

[98] See, for example, James R. Garfield to Frederick V. Stevens, Committee on Interstate Commerce (House), May 23, 1908 in memorandum prepared by Roosevelt, May 26, 1908 (TR).

Sensing the confusion of these procedures, Roosevelt, in a message to Congress, proposed that the Department of the Interior should exercise authority over water power permits in navigable streams.[99] Secretary Taft was appalled at the president's maneuvering. To permit a subordinate, Pinchot, in one department to establish a cabinet officer's policy in another, he later argued, produced only administrative disorganization and demoralization. These experiences played no little part in forming Taft's hostile attitude toward Pinchot and Garfield, which came to the fore when the Secretary of War became president in 1909.[100]

In 1908 Taft bided his time on the water power question, but, when he became president, he reversed his predecessor's policies and approved franchises which required neither a limited permit nor compensation. Early in 1911, however, Taft set out to appease some of Roosevelt's former friends, who now threatened to oppose his renomination in 1912. He appointed Henry L. Stimson, an active conservationist and member of the Board of Directors of the National Conservation Association, as his new Secretary of War. Stimson did not share Pinchot's crusading zeal, but he agreed with his water power policy, and was enthusiastic over the possibilities of using revenue from water power to construct multiple-purpose river works.[101]

The federal administration, Stimson argued, could constitutionally charge a rental for water power produced at privately owned dams in navigable streams, because such installations aided navigation and the federal government could clearly promote navigation. Heretofore the private power companies, Taft, and the Corps had distinguished sharply between public dams to improve navigation

[99] 60th Congress, 1st Session, *Congressional Record*, 4698–4699.

[100] Taft to Helen Taft, Oct. 3, 1909 (WHT, Presidential Series 3, folder 1).

[101] Henry L. Stimson and McGeorge Bundy, *On Active Service in Peace and War* (New York, 1948), 41–43. In his views on water power policy, Stimson leaned heavily on the advice of Felix Frankfurter, who had been his assistant as District Attorney for the Southern District of New York, and who was now Chief Law Officer of the Bureau of Insular Affairs in the War Department. See Stimson to Senator Knute Nelson, July 25, 1912 (RG #77, #49678/37).

and private dams for power production. In the first case, they argued, the federal government could charge for power, but in the second it could not.[102] This sharp distinction, Stimson argued, overlooked the intimate connection between navigation and water power, that in most cases a private dam for water power clearly affected navigation and therefore came under the power of Congress to regulate interstate commerce.[103] Taft and the Corps looked upon power dams not as opportunities for expanding navigation, but as obstructions to it.[104] Stimson, on the other hand, linked the two in the hope that their joint development could create benefits which separate construction could never produce.

Throughout his two years as Secretary of War, Stimson tried to secure legislation that would provide for federal rentals. The General Dam Act of 1910, he felt, did not grant him the power to impose a fee by administrative action. Although Taft did not agree with his water power views, Stimson persuaded the President in August 1912 to veto the Coosa, Alabama dam bill because it did not contain provisions for a rental.[105] But he could not as easily convince Congress. Stimson worked closely with private power firms, especially the Stone and Webster company, to devise a permit for a specific project, including rentals, which private and public groups could then jointly back in Congress. Such a measure would serve as a precedent for more general legislation.[106] This plan, involving construction of a dam on the Connecticut River, progressed rapidly, but

[102] Taft to Pinchot, Oct. 12, 1909 (GP #120).

[103] Stimson to Rep. William C. Adamson, July 30, 1912 (RG #77, #49678/43).

[104] See Taft's opinion as Secretary of War in the case of the Economy Light and Power Company, in 60th Congress, 2nd Session, *Congressional Record*, 2607.

[105] 62nd Congress, 2nd Session, *Sen. Doc. 949*. Stimson wrote this veto message. Taft did not agree with it, but followed Stimson's lead to avoid a cabinet row. See statement of Senator John H. Bankhead, 64th Congress, 1st Session, *Congressional Record*, 2580.

[106] Kerwin, *Water-Power Legislation*, 158; Stimson to Adamson, August 14, 1912 (RG #77, #49678/59); Pinchot to Allen Chamberlain, Jan. 10, 1913 (GP #1822).

met crucial opposition in Congress. Senator Bankhead of Alabama had sponsored the Coosa bill. When it failed to pass over the President's veto, he turned on the administration in February 1913 and personally blocked the Connecticut River proposal.[107]

By the spring of 1913 the water power struggle in Congress had deadlocked. Congress had checked the Roosevelt policy, but conservationists had also blocked perpetual, free grants to private corporations. This stalemate halted all water power development on the navigable streams at a time when it proceeded rapidly on the public lands. After a decade of legislative struggle the compromise Water Power Act of 1920 once more permitted development. This Act dealt solely with water power. It failed to realize the hopes of conservationists that power development would go forward as an integral part of a multiple-purpose river program.

[107] Harry A. Slattery to Mark Sullivan, Dec. 24, 1912 (GP #1820); Kerwin, *Water-Power Legislation*, 160.

Chapter VII

The Conservation Crusade

Until the spring of 1908 the Roosevelt administration's natural resource program moved forward without interruption. It had expanded from piecemeal irrigation, forest, range, and mineral measures to the most comprehensive water development program yet devised. New obstacles, however, appeared on the horizon. On the one hand, executive or administration action could expedite little of the expanded water program; on the other, Congress hesitated even to continue the Inland Waterways Commission, let alone enact its recommendations. Moreover, powerful influences in the War Department, among them Roosevelt's chosen successor in the White House, William Howard Taft, so disapproved of these new ventures that the President took special measures to overcome their opposition. Despite these obstacles, Pinchot and Garfield pressed for an even more comprehensive policy for the "conservation" of all natural resources. This program progressed during 1908 and 1909, but in the face of ever-increasing hostility from Congress, from the new President, and from members of the executive departments. Faced with this opposition, those in the vanguard of the Roosevelt resource movement turned more and more to the general public for support, and unleashed a veritable crusade of enthusiasm for conservation. In this fashion a movement peculiar to federal scientists and planners became deeply rooted in the minds of the public at large.

The Gospel of Efficient Planning

By the time of the great crusade, the threads of resource policy had become interwoven in a single coherent approach: the use of "foresight and restraint in the exploitation of the physical sources of wealth as necessary for the perpetuity of civilization, and the welfare of present and future generations."[1] This, in 1908, was conservation, a concept which formerly had referred to reservoir storage of flood waters and controlled grazing on the Western range, but had now come to connote efficiency in the development and use of all resources. Richard T. Ely, for example, Professor of Economics at the University of Wisconsin, defined conservation as "the preservation in unimpaired efficiency of the resources of the earth. . . ."[2] Indeed, for many the term implied the need for efficiency in social and moral affairs as well as in economic life. Like Theodore Roosevelt, they looked upon conservation as a major step in the progress of civilization; it would bring conscious foresight and intelligence into the direction of all human affairs.[3]

The conservation movement was closely connected with other organizations which attempted to promote efficiency. Leaders of the Roosevelt administration, for example, maintained close contact with the four major engineering societies, the American Society of Civil Engineers, the American Society of Mechanical Engineers, the American Institute of Electrical Engineers, and the American Institute of Mining Engineers. The societies spearheaded the drive for efficiency. They argued that the potential electrical energy in the nation's streams should be harnessed to lessen the drain on exhaustible coal supplies. They promoted more efficient mining methods and the utilization of by-products in the iron and steel industry, and pressed upon manufacturers the need for scientific management in the plant. Professional engineers felt a close kinship with the scientific and technological spirit of the Roosevelt administration. They took a keen interest in resource problems, publicized conservation affairs in their journals, and defended the administration from attack. They applauded heartily when the President

[1] Philip P. Wells, "Conservation of Natural Resources," MSS in GP #1672.

[2] Richard T. Ely and others, *The Foundations of National Prosperity* (New York, 1918), 3.

[3] Theodore Roosevelt to Kermit Roosevelt, May 17, 1908 (TR).

sought to bring order, efficiency, and business methods into government.

WJ McGee emerged as the prophet of the new world which conscious purpose, science, and human reason could create out of the chaos of a laissez-faire economy where short-run individual interest provided no thought for the morrow. McGee was the key figure in disseminating the expanding concepts of the conservation movement, and in so doing he exhibited much of the passion for directed social planning held by his contemporary, the sociologist, Lester Ward. Again and again at resource conclaves McGee emphasized his faith in applied knowledge: "the course of nature has come to be investigated in order that it may be redirected along lines contributing to human welfare. . . ."[4] The conservation movement he considered a momentous step in human progress because it involved "a conscious and purposeful entering into control over nature, through the natural resources, for the direct benefit of mankind."[5] The Inland Waterways Commission, with its vision of massive river development, he maintained, involved the "highest application of applied anthropology thus far attempted."[6] In his speeches and writings McGee formalized the spirit of efficient planning implicit in the policies of Roosevelt resource administrators.

Concern for the elimination of natural resource waste contributed to a wider gospel of efficiency in every phase of human life. Joseph N. Teal, a prominent Oregon conservationist, for example, declared, "I hope that the time will come when efficiency in all directions will be given consideration. When that time comes we will begin to get our money's worth for what we spend."[7] The ultimate goal, argued McGee, was absolute efficiency. "The perfect machine," he wrote, "is the fruit of the ages. . . ." The editor of an engineering journal which heartily supported resource conservation proclaimed:

[4] McGee, "Water as a Resource."

[5] *Proceedings, National Conservation Congress, 1911*, 184.

[6] McGee to Dr. George Grant MacCurday, Apr. 8, 1907 (WJM, Miscellaneous Letterbook #3).

[7] Joseph N. Teal to Francis G. Newlands, Feb. 28, 1916 (FGN, Waterways, 1916, 1).

"The Millennium will have been reached when humanity shall have learned to eliminate all useless waste. . . . When humanity shall have learned to apply the common sense and scientific rules of efficiency to the care of body and mind and the labors of body and mind, then indeed will we be nearing the condition of perfect."[8] And Theodore Roosevelt declared at the Conference of Governors in Washington in May 1908, "Finally, let us remember that the conservation of natural resources . . . is yet but part of another and greater problem . . . the problem of national efficiency, the patriotic duty of insuring the safety and continuance of the Nation."[9]

The Roosevelt administration pursued this concern for efficiency into its own executive departments. Emphasizing training and ability, the new resource agencies—such as the Geological Survey, the Reclamation Service, and the Forest Service—selected officials through competitive examinations. In 1905 the President appointed the Keep Commission on Departmental Methods to unearth executive inefficiencies and to formulate policies for improved administration.[10] In explaining to Congress his purpose in establishing the Commission, the President declared: "There is every reason why our executive governmental machinery should be at least as well planned, economical, and efficient as the best machinery of the great business organizations."[11] Even though Congress did not appropriate the $25,000 which Roosevelt requested for its work, the Commission helped to reduce government printing expenses and to improve methods for testing and purchasing government supplies. It completed a thorough investigation and report on the administration of the Department of the Interior, and it helped to stimulate

[8] *Cassier's Monthly* (July 1913), 44, 1.

[9] *Proceedings of a Conference of Governors in the White House, Washington, D.C.*, May 13–15, 1908 (Washington, D.C., 1909), 12.

[10] For the Keep Commission see Henry Beach Needham, "New Business Methods in National Administration," *World To-day* (Dec. 1905), IX, 1332–1339; C. H. Forbes-Lindsay, "New Business Standards at Washington—Work of the Keep Commission," *American Review of Reviews* (Feb. 1908), XXXVII, 190–195; Pinchot, *Breaking New Ground*, 296–299.

[11] *A Compilation of Messages and Papers of the Presidents*, 6989.

among federal employees a greater interest in efficient administration.[12]

The President, Pinchot, and Garfield carried their interest in efficiency into a variety of fields other than natural resources. They emphasized, for example, the value of large-scale business organization, and warned that anti-trust action might impair increased production. They had no quarrel with bigness as such. Instead, in a manner similar to Thorstein Veblen, they distinguished between the efficient production engineer and the adventurous profit-taker who sacrificed a wider distribution of cheaper goods for speculative investment returns. Such profits, declared Pinchot, were "a toll levied on the cost of living through special privilege." Speculative profit-taking produced as much waste as did competitive exploitation of natural resources. In either case conservationists strove to encourage the greatest possible production of material goods at the lowest cost. Just as planning must replace competition so that manufacturers could produce with less waste, so regulation must prevent financial freebooters from destroying industrial efficiency.

The efficiency expert also feared that the organized labor movement would hamper more rational production. One could not neglect the conditions of workingmen, so the argument ran, for labor unrest could disrupt industrial efficiency. But to deal with the problem through unionization would be self-defeating. The workingman was concerned with wages and working conditions rather than with expanded production at lower cost, and the labor union would readily sacrifice efficiency for its own interest. The efficiency expert tended to reject the entire idea of class consciousness as potentially disruptive of that order essential for smooth industrial operations. He tackled the working-class problem not from a concern for human rights as such, or even for a more equitable distribution of wealth, but from a judgment as to what conditions were required for industrial efficiency. He therefore approved of public

[12] The results of the work of the Keep Commission are described briefly in Pinchot, *Breaking New Ground*, 296–299; the entire work of the Commission may be followed in manuscript form in GP #1933–#1935.

welfare programs which would alleviate discontent without impairing his control. Of all such reforms the engineer most enthusiastically supported measures to reduce industrial accidents, which would not only protect the laboring man, but would also increase his usefulness as a producer.

The apostles of the gospel of efficiency subordinated the aesthetic to the utilitarian. Preservation of natural scenery and historic sites, in their scheme of things, remained subordinate to increasing industrial productivity. Forestry, Pinchot argued, did not concern planting roadside trees, parks, or gardens, but involved scientific, sustained-yield timber management.[13] During a conservation meeting early in 1908, the president of the American Society of Civil Engineers expressed the prevailing attitude of the members of his profession on this question. Lord Kelvin, he declared, was once asked how water power development at Niagara Falls would affect its natural beauty; "His reply was that of a true engineer: 'What has that got to do with it? I consider it almost an international crime that so much energy has been allowed to go to waste.' " Archeologists, the Society's president complained, had argued that irrigation works on the Nile would inundate ancient Egyptian ruins, but "engineers," he assured his audience, "will naturally consign all such archaic questions to the oblivion of the past, and concern themselves with that which confers the greatest good upon the greatest number."[14]

The Great Inventory

This expanded philosophy of efficient planning found tangible expression in the Governors' Conference in 1908 and in the subsequent inventory of natural resources undertaken by the National Conservation Commission. These events were foreshadowed in Roosevelt's original letter appointing the members of the Inland Waterways Commission, when he asked them to "consider the

[13] Pinchot, *Breaking New Ground*, 71.

[14] Address by Charles Macdonald, president, American Society of Civil Engineers, in *Proceedings, American Society of Civil Engineers* (Aug. 1908), 34, 254.

relations of the streams to the use of all the great permanent natural resources and their conservation for the making and maintenance of prosperous homes."[15] Yet at this time the administration had not fully formulated either the concept of conservation or the comprehensive measures to implement it. In fact, the proposal for a Governors' Conference came originally in May 1907 from the Inland Waterways Commission and was limited to the "purpose of ascertaining through the State Governors and other public men the present requirements of the country with respect to the improvement of waterways and the development of water resources."[16] The original press release announcing the Commission's request also placed major stress upon water problems; it suggested a convention to consider "the conservation and utilization of waters and other natural resources. . . ."[17]

Under the influence of McGee the administration shifted the emphasis of the Governors' Conference to the conservation of all natural resources. In a formal letter in which McGee, on behalf of the Inland Waterways Commission, recommended that Roosevelt call the conference, he stressed conservation as a single over-all problem. Roosevelt responded to this invitation with an equal emphasis on the singleness and gravity of the task:

> As I have said elsewhere, the conservation of natural resources is the fundamental problem. Unless we solve that problem it will avail us little to solve all others. . . . As a preliminary step, the Inland Waterways Commission has asked me to call a conference on the conservation of natural resources, including, of course, the streams, to meet in Washington during the coming winter. I shall accordingly call such a conference. It ought to be among the most important gatherings in our history, for none have had a more vital question to consider.[18]

Moreover, Roosevelt's letter of appointment to the Governors' Conference, written in November of 1907, echoed this shift in emphasis;

[15] Roosevelt to members of the Inland Waterways Commission, Mar. 14, 1907 (TR).

[16] IWC, *Minutes*, May 20, 1907.

[17] *Ibid.*, May 21, 1907.

[18] Pinchot, *Breaking New Ground*, 345.

he spoke of the "conservation of natural resources," and only peripherally mentioned water development.[19]

Meeting in Washington from May 13 to 15, 1908, the Governors' Conference included not only state executives and their aides, but also representatives from seventy national organizations. In addition, the President invited five men to represent the public, William Jennings Bryan, Andrew Carnegie, John Mitchell, James J. Hill, and former President Cleveland (who could not attend because of illness). Although those in charge of the Conference allotted time to the governors to present speeches, most of the proceedings consisted of papers on conservation problems, read by experts, but written primarily by Pinchot and McGee. Following their lead, the governors at the end of the Conference issued a statement reiterating the great need for conservation and recommending that the President appoint a National Conservation Commission to undertake a national inventory of all natural resources.[20]

The proposal for a national resources inventory came initially from the engineering societies. These organizations had maintained contact with the Roosevelt administration for a number of years through a National Advisory Board on Fuels and Structural Materials, established to supervise the testing of the quality of fuels and construction material used by the federal government.[21] When invited to send representatives to the Governors' Conference, they immediately appointed delegates, organized conservation committees, and held meetings to arouse interest among engineers.[22] In March 1908 Charles Whiting Baker, editor of *Engineering News*, suggested the national inventory project to Pinchot as a proposal to

[19] Roosevelt to the Governors, Nov. 11, 1907 (TR).

[20] The Conference can be followed in *Proceedings, Conference of Governors, 1908*.

[21] Roosevelt to Richard L. Humphrey, Mar. 2, 1906; Roosevelt to Taft, Apr. 20, 1906; Roosevelt to Charles B. Dudley, Mar. 12, 1907 (all in TR).

[22] See resolution of the American Society of Civil Engineers, in *Proceedings, American Society of Civil Engineers* (May 1908), 34, 200; Roosevelt to President, American Society of Civil Engineers, Jan. 14, 1908 (TR); Roosevelt to Calvin W. Rice, Apr. 13, 1908 (TR).

be sponsored by the four engineering societies.[23] The Chief Forester and McGee immediately took up the suggestion and presented it to the Conference in the form of a resolution which the governors adopted.[24]

To many conservationists this inventory merely applied sound business principles to the entire nation. In his first report on the Commission's work, Roosevelt echoed this viewpoint:

All we are asking . . . is that the National Government shall proceed as a private business man would, as a matter of course, proceed. He will regularly take account of stock, so that he may know just where he stands. If you find that he does not, that he does not know how his outgo corresponds with his income, you will be afraid to trade with him. The same measure of prudence demanded from him as an individual, the same measure of foresight demanded from him as an individual, are demanded from us as a nation. Unfortunately, nations have been slow to profit by the example of every individual among them who makes a success of his business.[25]

Businessmen who became active in the movement voiced similar attitudes. One executive described resource exploitation as merely a "Get Rich Quick" scheme, a means of drawing on national capital to pay immediate dividends, for which laws in many states held a man personally liable in private affairs. The national inventory, he argued, would merely measure the nation's capital to determine if it were being sacrificed for greater yearly returns.[26]

The Roosevelt administration hoped that the National Conservation Commission would consist of experts. In fact, Pinchot preferred that Congress not take up the governors' resolution for fear that the resulting body would contain Senate and House members exclusively, and would hamper rather than help executive action during the remainder of 1908.[27] Congress did not act, and, after its

[23] Charles Whiting Baker to Pinchot, Mar. 24, 1908 (GP #2137).

[24] Pinchot to Baker, Apr. 4, 1908 (GP #2137).

[25] *Conservation* (Jan. 1909), 15, 5.

[26] Wallace D. Simmons, "Our Resources as the Basis for Business," *Proceedings, National Conservation Congress, 1910*, 257–263.

[27] Pinchot to Elbert F. Baldwin, May 27, 1908 (GP #2137).

adjournment, Roosevelt appointed the Commission. He chose its personnel carefully, placing in prominent positions congressmen friendly to his conservation program, but taking care that leaders of the federal resource departments would actually direct the work and formulate conclusions. As chairmen of each of the Commission's four sections—Waters, Forests, Lands, and Minerals—the President chose either a United States Senator or Representative. But, for the sections' secretaries, the members who carried the brunt of the work, he selected four tried and true conservation stalwarts, WJ McGee, Overton W. Price of the Forest Service, George Woodruff of the Department of the Interior, and Joseph A. Holmes of the Geological Survey. Since the Commission could rarely meet as a group, Roosevelt appointed an executive committee to direct its activities, composed of the four section leaders, their secretaries, and Chief Forester Pinchot as chairman.[28]

Hampered because Congress had not authorized funds for its work, the Commission could undertake no new investigations, but it could draw together material already collected in one over-all account of the nation's resources.[29] The President directed the heads of federal scientific bureaus to furnish information which the Commission requested, and, in some cases, he ordered that special men be detailed to compile it. Roosevelt assigned Henry Gannett, Geographer of the United States Geological Survey, to supervise compilation of the data, while the four section secretaries directed the preparation of the report.[30] Early in December 1908 the full Commission received and approved the document, and a few days later a second meeting of governors also accepted it.[31]

[28] Roosevelt to members of the National Conservation Commission, June 8, 1908 (TR).

[29] Pinchot describes the work of the Commission in *Breaking New Ground*, 355–360.

[30] Roosevelt to Dr. Milton Whitney and others, July 2, 1908; Roosevelt to G. W. W. Hanger and others, July 10, 1908; Roosevelt to Frank Pierce, July 2, 1908 (all in TR).

[31] "Proceedings of the Joint Conservation Congress" in 60th Congress, 2nd Session, *Sen. Doc. 276, Report of the National Conservation Commission*, 123–267.

In this three-volume report the National Conservation Commission compiled the most extensive inventory of the nation's natural resources then available. Each section reported on the supply of resources, their rate of use, and the probable date of their exhaustion. The Lands subdivision concentrated on the public domain; the Waters section continued the work of the Inland Waterways Commission; the Minerals group dealt with the national mineral supply; and the Forest committee revealed that annual timber cutting exceeded annual growth by 250 per cent. The Commission outlined specific measures which the federal government should undertake: leasing of the public coal and grazing lands, creation of Eastern national forests, development of flood waters, prevention of soil erosion, classification of existing public lands according to their highest use, and repeal of out-of-date land laws. Few of these suggestions were new; they added little, if anything, to measures which Pinchot, Garfield, Newell, and Roosevelt had long advocated. The Commission merely brought together these past proposals, giving them the form of a recommendation for a comprehensive policy.[32]

The second Conference of Governors in December 1908 suggested that the National Conservation Commission continue its work and that the states establish similar organizations. To guide and coordinate these activities the governors appointed a Joint Committee on Conservation Between State and Nation. By the middle of 1909, over forty-one states had created commissions, each one of which investigated the extent and rate of use of one or more of the state's resources. Many of these groups merely brought together previously formed geological surveys, game bureaus, and resource development agencies into a single conservation office. Others, however, were newly organized as a result of the enthusiasm for conservation created by the national movement.[33]

The National Conservation Commission, however, soon faced opposition from Congress which led to its demise. In February 1909

[32] 60th Congress, 2nd Session, *Report of the National Conservation Commission*, 39–111. A brief summary of the report and its recommendations is in Robbins, *Our Landed Heritage*, 356–362.

[33] Pinchot to Paris Gibson, June 5, 1909 (GP#120).

Roosevelt requested that Congress appropriate $25,000 for the Commission's expenses. Although Senator Knute Nelson of Minnesota, chairman of the Commission's section on lands, introduced a bill for this purpose, it died in committee.[34] President Taft asked Congress to continue the National Conservation Commission by special act, and Senator Newlands introduced bills to carry out this request, but to no avail.[35] The Commission's work was at an end.

Congress and the Executive

The National Conservation Commission brought to a head an increasingly strained relation between Congress and the Roosevelt administration over resource policy. This controversy went to the heart of the conservation idea itself, the view that experts alone should make resource decisions. Congress, so the argument ran, could not deal with such complex questions; it was better able to fulfill the immediate desires of its local constituents than to carry out a rational resource program in the light of scientific facts. Moreover, the great number of differing opinions among congressmen rendered them incapable of acting with the dispatch essential to operate large and intricate programs efficiently. President Roosevelt expressed these views when he wrote to a friend, "I am afraid all modern legislative bodies tend to show their incapacity to meet the new and complex needs of the times."[36] With more dramatic effect, Joseph N. Teal, chairman of the Oregon Conservation Commission, agreed:

> The great difficulty in this country, and I presume in all democracies, lies in the fact that . . . the views of experts are of little value, and that a slapdash-haphazard way of going at anything will produce results. . . . it reminds me of when I was riding after cattle, and a bunch of

[34] Ise, *U. S. Forest Policy*, 153, contains a brief account of these developments.

[35] 61st Congress, 2nd Session, *Sen. Rept. 826*; 61st Congress, 2nd Session, *S 3719*. Pinchot and others tried to persuade Taft to continue the commission through executive action, but failed; see Pinchot, *Breaking New Ground*, 406–407.

[36] Roosevelt to Dan T. Moore, Jan. 9, 1909 (TR).

cattle would get to "milling," and they could go round and round for hours, starting nowhere and getting nowhere.[37]

Dissatisfaction with the inability of Congress to act efficiently cropped up continually during Roosevelt's terms of office. In 1905, speaking of the need for a more efficient federal administration, the President declared: "to make it so is a task of complex detail and essentially executive in its nature; probably no legislative body, no matter how wise and able, could undertake it with reasonable prospect of success."[38] Senator Newlands criticized the inefficient logrolling methods used by Congress to undertake rivers and harbors improvements. He threw up his hands in disgust when congressmen refused to adopt a system of priorities in public works projects. Editors of engineering magazines, equally contemptuous of the congressional waterway program, proposed that an independent commission of experts plan and execute rivers and harbors developments.[39]

To enable them to act when Congress did not, Roosevelt, Pinchot, and Garfield devised new administrative concepts and practices. The executive is the steward of the public welfare, Roosevelt declared.[40] As direct representative of the people at large, he has a responsibility to guide the public and Congress toward that resource policy best for the entire country. The President could, for example, justifiably appoint commissions without congressional approval to investigate problems which he felt the nation should tackle. Roosevelt named no fewer than seven such commissions to report on the public lands, the conditions of country life, the inland waterways, scientific work in the federal departments, administrative efficiency, fuels and structural materials, and national conservation problems.[41] Often he

[37] Joseph N. Teal to Francis G. Newlands, Feb. 28, 1916 (FGN, Waterways, 1916, 1).

[38] *A Compilation of Messages and Papers of the Presidents*, 6989–6990.

[39] 60th Congress, 1st Session, *Congressional Record*, 395–396; *Engineering News* (Mar. 28, 1907), 57, 355; (Apr. 4, 1907), 57, 381.

[40] Roosevelt, *Autobiography* (New York, 1913), 405–406.

[41] Roosevelt, *Autobiography*, 365–369; see also Pinchot, *Breaking New Ground*, 240–243, 296–299, 340–344.

established these groups without specific congressional authorization, and thereby incurred the hostility of the lawmakers.

When Congress did not specifically approve his investigating commissions, the President was hard pressed to devise novel methods of financing their activities. Officials already employed by the federal government, for example, carried out the work of the National Conservation Commission; the Russell Sage Foundation paid the expenses of the Country Life Commission.[42] The Governors' Conference of May 1908 was financed largely from Pinchot's private income.[43] The administration also depended on nongovernmental funds to publish the findings of several of the commissions. The members themselves financed the preliminary report of the Inland Waterways Commission (Pinchot paid Brigadier-General Mackenzie's share of the costs),[44] and the Spokane, Washington, Chamber of Commerce published the recommendations of the Country Life Commission.[45] Congress, however, printed the final report of the National Conservation Commission as a public document.

The "new and complex needs of the times" required greater executive influence in shaping public policy as well as expert fact-finding commissions. Congress, so conservationists argued, should establish the general framework of policy and leave its details to the executive branch of the government. Senator Francis G. Newlands had in mind such a division of labor when he provided for wide executive discretion in the Reclamation Act of 1902. Congress should supervise, criticize, examine, and require frequent reports, he reasoned; it should exercise general supervision, but not burden a resource program with too detailed control.[46] Newlands emphasized to a friend the potentialities for executive action which his measure provided:

Under this Bill, thorough demonstration will be given of the efficiency

[42] Pinchot, *Breaking New Ground*, 341.

[43] Pinchot to Prof. Irving Fisher, May 6, 1908 (GP #1936).

[44] McGee to members of the Inland Waterways Commission (except Bankhead and Mackenzie), nd, but prior to June 17, 1908 (GP #2137).

[45] Pinchot, *Breaking New Ground*, 343.

[46] 60th Congress, 1st Session, *Congressional Record*, 395.

of the Government itself in taking hold of public utilities. If the work is conducted efficiently and honestly . . . it will demonstrate the capacity of Government for such enterprises. No legislative Act in the history of this Government is so full of suggestions as to the nationalization of great projects as this. Under it the Geological Survey has a free hand in the investigation of irrigation possibilities and the construction of irrigation works; their maintenance for the common good; the reservation of the public lands for actual settlers and homemakers; the prevention of monopoly.[47]

Newlands' Reclamation Act entailed precisely the division of responsibility between Congress and the executive which the conservationists desired.

To justify similar discretion in the administration of other resources, Roosevelt, Pinchot, and Garfield expanded the interpretation of laws such as the Forest Administration Act of 1897 and the Right-of-Way Act of 1901, and relied heavily upon their law officers to discover legal justification for new policies.[48] Pinchot's second law officer, Philip P. Wells, came to the Forest Service after he had become "disgusted with the intellectual method and narrowness of lawyers and judges in general and with the enormous difficulty of applying the existing legal system to social progress in any effective and comprehensive manner." Wells marveled at the way in which the logic of his predecessor, George Woodruff, had forced its way "from indisputable principles to inevitable conclusions with small regard for precedents to make the intermediate path."[49]

Greater autonomy in forming policy frequently depended upon financial independence from Congress. Senator Newlands fully recognized this fact when he provided in the Reclamation Act that the Secretary of the Interior be free to disburse the reclamation fund as he saw fit, and when he requested in his larger waterway bill that the river development commission be permitted to spend a lump-sum appropriation as it desired. Pinchot tried to follow the same

[47] Newlands to Dent H. Robert, Sept. 17, 1904 (FGN, Letters).

[48] Pinchot, *Breaking New Ground*, 417–418; Roosevelt, *Autobiography*, 357–358.

[49] Wells, "Personal History" (GP #1671).

principle in establishing and attempting to make permanent a "forestry fund" free from congressional influence. But the lawmakers did not relish these persistent attempts to evade their power of the purse. During the first five years of Roosevelt's presidency they threatened to retaliate, but not until 1906 did they openly defy the administration. In the Sundry Civil Bill of that year, the House of Representatives stripped the Geological Survey of all funds for hydrographic investigations. Representative Crumpaker of Indiana, the sponsor of this move, pointed out that no law had granted the Survey authority to undertake such a task, and that provisions for it did not even appear in the organic law creating the Survey. Coming to the defense of that department, others argued that for a dozen years Congress had granted similar requests without question. Yet Crumpaker succeeded in his move. The House deleted the funds, the Senate restored them, and in conference the two compromised on $150,000, a cut of $50,000 from the previous appropriation. The following year Congress pared the sum still further to $100,000.[50]

In 1907 the storm broke. When Pinchot proposed that the independent forestry fund be retained until Congress acted otherwise, the administration's opponents were up in arms. Both the national forest area and its revenue had increased each year. In fiscal year 1905–6 funds derived from the forests had risen to $800,000, almost to the amount, $875,000, which Congress had appropriated for the Forest Service. If Congress adopted his proposal, Pinchot argued, it could soon cut the regular appropriation to $400,000 and eventually abolish it completely. This argument did not impress the lawmakers. Besides Pinchot's revenue measure, the Sundry Civil Bill also contained an amendment providing for the lease of public domain grazing lands by the Department of Agriculture, an innovation which would further increase Forest Service income. In its wrath Congress not only rejected these propositions but agreed to the amendment demanded by a few Western senators that prohibited the president from creating new forest reserves in the six

[50] 59th Congress, 1st Session, *Congressional Record*, 8423–8432, 2537; 69th Congress, 1st Session, *House Rept. 1500, Sen. Rept. 265*.

northwestern states of Oregon, Washington, Idaho, Montana, Wyoming, and Colorado.[51]

Despite this reverse, the administration continued to try the patience of Congress and the lawmakers grew more fearful of the conservationists' desire for independence. In the spring of 1908 Congress refused to appropriate funds to continue the work of the Inland Waterways Commission. Immediately after Congress adjourned, Roosevelt reappointed the Commission as a section of the more extensive National Conservation Commission which he established also without legislative sanction. Those were the last straws. In February 1909, when Roosevelt sought funds for the work of the National Conservation Commission, Congress not only denied his request but also approved the Tawney amendment to the Sundry Civil Bill, prohibiting any federal administrative official from aiding the work of any executive commission not authorized by Congress. This was the final answer of Congress to an administration which the lawmakers thought had gone too far in assuming legislative powers.[52]

The Bid for Popular Support

By 1908–9 Congress had blocked most of that part of the conservation program which depended on legislative approval. In response, the administration took its case to the public to create a national sentiment in behalf of its resource program which Congress could not ignore. From the first years of his federal forestry work, Pinchot had recognized the lack of popular enthusiasm for scientific forest management and the need to educate the public so that the federal forest program could withstand congressional attack. The Forestry Bureau and the Forest Service established contacts with many newspapers; they issued regular press releases recounting their work and recommending expansion of their programs.[53] To arouse popular support when an apathetic Congress threatened to block the transfer bill in 1905, Pinchot stepped up his press releases and

[51] This episode can be followed best in Peffer, *Public Domain*, 90–98.

[52] Ise, *U. S. Forest Policy*, 152.

[53] Peffer, *Public Domain*, 66–69.

called a Forest Congress to meet in Washington. Although portrayed as a great landmark in the history of scientific forestry, that meeting was actually a technique to persuade the lawmakers to pass the transfer measure. It aroused widespread public interest in forestry and undoubtedly helped to realize its intended objective.[54]

These devices, although valuable in securing the transfer bill, did not create a sustained popular support which could help administration leaders persuade Congress to pass other measures. In the winter of 1906–7, however, the wave of sentiment for improvement of the inland waterways provided conservationists with a wonderful opportunity to rally the public behind its entire resource program. Badly in need of support after congressional opposition had risen to new heights in the spring of 1907, the administration made every effort to exploit this enthusiasm for water development. The Inland Waterways Commission, for example, established close ties with the Lakes-to-the-Gulf Waterway Association. Upon that organization's invitation, Pinchot and Roosevelt journeyed down the river from Keokuk, Iowa to Memphis, Tennessee, where, at the Association's second annual convention, the President publicly endorsed the waterways movement.[55]

Through the Governors' Conference the administration hoped to channel this enthusiasm for waterways into support for its entire resource program. Pinchot and McGee especially visualized its political possibilities, and developed plans for the conference in such a way as to exploit the meeting fully. To include not merely the governors, but also representatives from all national organizations which had shown interest in conservation, they postponed the date of meeting from November 1907 to May 1908 and, in the meantime, lengthened the list of invitations.[56] But Pinchot and

[54] *Proceedings of the American Forest Congress, 1905* (Washington, D.C., 1905); Pinchot, *Breaking New Ground*, 254–256.

[55] William F. Saunders, "The President's Mississippi Journey," *American Review of Reviews* (Oct. 1907), XXXVI, 456–460; Pinchot, *Breaking New* Ground, 329–330.

[56] Roosevelt to Burton, Aug. 21, 1907 (TR); Roosevelt to Governors, Nov. 11, 1907 (TR).

McGee carried their political enthusiasm too far for the President. They sent to him for his signature invitations to officials of a number of patriotic societies. Indignantly demanding that the list of delegates be held within bounds, Roosevelt returned the letters unsigned.[57]

Meeting from May 13 to 15, 1908, this "Governors' Conference" was well controlled by administration leaders. McGee and Pinchot drew up its agenda, wrote many of the speeches, and planned its action.[58] Opposition governors from the West received time to reply, and they denounced the grazing fees and demanded that Congress cede the public lands to the states. But their protests were quietly shelved. Although delegates representing so many different sectional and political interests could hardly agree on positive action, Pinchot and McGee felt that the Conference should produce some specific recommendations in order to fix the gathering more clearly in the public mind. The commission to inventory all natural resources was the answer, a safe proposal which could antagonize no one, and which all could support. Moreover, although the governors would not rally behind the measures which the administration hoped Congress would adopt, a Commission, whose members the conservationists could select, might do so. The strategy succeeded. To the public the great inventory was not only the culmination of the conservation idea, but the singular and triumphant work of the Governors' Conference. And the National Conservation Commission gave far more unified support to the administration's policies than did the governors.

The larger strategy of using the Governors' Conference as a "big stick" to persuade Congress to act did not succeed. Instead, in the Tawney amendment, the lawmakers indicated that the administration's actions only infuriated them further, and forced Pinchot and Garfield to turn to other means of arousing public sentiment. Already realizing the increasing political weakness of the conserva-

[57] Roosevelt to McGee, Mar. 20, 1908 (TR).

[58] Pinchot, *Breaking New Ground*, 344–355; the records of the Governors' Conference in the Pinchot papers confirm this statement.

tion program, the Chief Forester was at work forming a national organization that would build a fire under Congress.[59] The need for such an organization increased in 1909 when the new president, William Howard Taft, refused to continue the work of the Commission in the face of the crippling Tawney amendment. During 1908 Pinchot had organized the Conservation League of America as a pressure group to support the administration's policies.[60] Composed primarily of waterway associations, the Conservation League gave way in the summer of 1909 to the National Conservation Association. This group, Pinchot hoped, would attract sufficiently wide popular following to break the deadlock over resource policy between Congress and the Roosevelt resource leaders.[61]

Conservation as a Moral Crusade

This bid for popular support brought into the conservation movement a new and somewhat disturbing influence. Heretofore, specific interest groups, such as waterway organizations, Western cattle associations, lumbermen's organizations, and national engineering societies had comprised the major political backing for the administration's resource policies. Concerned primarily with economic growth, they aided Pinchot and his friends because of a common interest in rational development. Those who came to the support of conservation in 1908 and 1909, however, were prone to look upon all commercial development as mere materialism, and upon conservation as an attempt to save resources from use rather than to use them wisely. The problem, to them, was moral rather than economic. An exclusively hardheaded economic proposition, therefore, became tinged with the enthusiasm of a religious crusade to save America from its materialistic enemies.

[59] Pinchot to Walter L. Fisher, Feb. 23, 1909 (GP #120).

[60] Pinchot to Charles R. Crane, June 4, 1908 (GP #2137); see folder, "Conservation League of America," in GP #120; *Conservation* (Oct. 1908), 14, 562.

[61] Pinchot to Fisher, Feb. 23, 1909; Pinchot to Charles W. Eliot, June 7, 1909; Pinchot to Curtis Guild, Jr., June 7, 1909; Pinchot to Eliot, Aug. 5, 1909 (all in GP #120); *Conservation* (Oct. 1909), 15, 640–641; (Feb. 1910), 16, 123–124.

This new interest in conservation, which dominated the movement between 1908 and 1910, came primarily from middle- and upper-income urban dwellers. Most of the new conservation organizations recruited their members and obtained their financial support from urbanites, and many of their leaders had been active in other types of urban reform. William Kent, Alfred N. Baker, and Walter L. Fisher, all prominent conservationists after 1908, had played important roles in the campaign of the Chicago Municipal Voters League for reform in the government of the Windy City. Women's organizations such as the General Federation of Women's Clubs and the Daughters of the American Revolution became especially enthusiastic about conservation. A leader of the National Conservation Association wrote that three women leaders in these two organizations had "done as much in the legislative field for conservation as any three men I know of."[62] The Daughters of the American Revolution maintained a special Committee on Conservation, of which Pinchot's mother, Mrs. James Pinchot, was chairman for several years. And Pinchot himself declared that the Daughters of the American Revolution "federated and organized spells only another name for the highest form of conservation, that of vital force and intellectual energy."[63]

Groups such as these viewed with alarm the way in which industrialism, in a short space of fifty years, had altered American society; they looked upon conservation as an antidote to changes they resisted. The organization of industry into combinations and labor into unions, they feared, threatened the traditional, independent, self-made man. Sprawling urban monstrosities were replacing sobriety, honesty, and hard work with disease, immorality, and squalor. Political decisions no longer came as a result of reasonable action by intelligent men, but involved a crude power struggle dominated by privileged wealth and effective city machines. Religious values and personal morality were giving way to secular life and group action. Everywhere one saw ugly urban centers, the

[62] Harry A. Slattery to George Kibbe Turner, May 26, 1913 (GP #1827).
[63] *Proceedings, National Conservation Congress, 1911*, 117.

conspicuous consumption of huge fortunes, and a headlong worship of the almighty dollar. Traditional American virtues, so these discontented argued, were on the point of extinction.[64]

Conservation, even though often in a vague and general way, symbolized the direct opposite of these ominous tendencies. It was oriented toward the countryside, toward nature and the eternal values inherent in nature, rather than toward the more artificial, materialistic, and socially unstable cities. The crusade of 1908–10, for example, attracted many people interested in parks and wilderness areas, in rural life, and in nonmaterialistic values. Industrialism and urbanization, to them, had spawned a "grovelling lust for material and commercial aggrandizement."[65] Urban dwellers had to turn elsewhere for the "regeneration of the spirit of man," and this they found in the wonders and beauties of nature which alone could "sustain life and make life worth sustaining."[66] In the latter part of the nineteenth century many organizations arose in the larger cities to campaign for more city parks, new recreation areas, the development of outdoor art, and the elimination of advertising billboards. In 1900 these groups joined to form the American League for Civic Improvement, and four years later this group became the American Civic Association. J. Horace McFarland, a civic leader and newspaper editor of Harrisburg, Pennsylvania, spearheaded the Association's activities and broadened its scope of action to campaign for state and national parks.[67] The General

[64] For a good description of this same frame of mind among California reformers see George E. Mowry, "The California Progressive and His Rationale: A Study in Middle Class Politics," *Mississippi Valley Historical Review* (July 1949), 36, 239–250.

[65] Mrs. Matthew T. Scott in *Proceedings, National Conservation Congress, 1910*, 272.

[66] Henry A. Baker in *Proceedings, National Conservation Congress, 1909*, 103.

[67] Charles Zueblin to Pinchot, May 9, 1902 (GP #81); Robert Shankland, *Steve Mather of the National Parks* (New York, 1951), 51–54. The American Civic Association was formed June 10, 1904 by a merger of the American Park and Outdoor Art Association and the American League for Civic Improvement.

Federation of Women's Clubs was particularly active in the forest and park movement. It organized campaigns to save the Minnesota National Forests, the Palisades Park in New York and New Jersey, and the Big Trees in California; it agitated effectively for the bill to establish the Appalachian National Forest.[68]

An interest in rural life, closely related to the concern for parks and wilderness areas, also helped to draw many urban enthusiasts into the conservation crusade. Urban leaders had long been attracted to the federal irrigation program, and had supported the National Irrigation Association which backed the work of the Bureau of Reclamation. They agreed especially with its emphasis on creating rural homes. They approved the sentiment of the Association's President, Thomas F. Walsh, when he wrote: "Man's inherent and ineradicable love for the soil is one of the strongest traits of human nature. . . . This is our natural taste, while the fascinations of town life are artificial. They do not satisfy our deeper feelings."[69] Fearing that mushrooming cities also threatened the United States with social disorder, many hoped to promote rural life as a stabilizing factor in society. They agreed with David Starr Jordan that "stability of national character comes from firmness of foothold in the soil."[70] Responding to this concern for rural life, President Roosevelt in 1908 appointed the Country Life Commission to investigate rural conditions and to recommend measures for their improvement.[71] Three years later, the third session of the National Conservation Congress, a loosely knit organization of state associations, was devoted exclusively to rural life problems. Henry Wallace, an Iowa agricultural leader and editor of *Wallace's Farmer*, was elected president of the Congress, and Liberty Hyde Bailey, the recognized leader of the Country Life Movement, took a prominent part in its work.[72]

[68] *Proceedings, National Conservation Congress, 1910*, 160–163.

[69] Thomas F. Walsh, "Humanitarian Aspect of National Irrigation," *Forestry and Irrigation* (Dec. 1902), 8, 505–509.

[70] *Maxwell's Talisman* (Sept. 1905) 5, 3.

[71] Pinchot, *Breaking New Ground*, 340–344.

[72] *Proceedings, National Conservation Congress, 1911.*

The common denominator which drew all these groups together and attracted them to the conservation movement was a feeling that a desire for material gain had become too prominent in America, and that other values should be stressed. A representative of a women's organization declared at a meeting of the National Conservation Congress: "We feel that it is for us, who are not wholly absorbed in business, to preserve ideals that are higher than business. . . ."[73] In greater contact with nature one could renew his spiritual life. "National Parks," declared one crusader, "represent opportunities for worship through which one comes to understand more fully certain of the attributes of nature and its Creator."[74] In a more general manner many felt that the conservation movement aroused Americans from their materialistic lethargy to fight for "higher ideals." Alfred L. Baker, a prominent Chicago conservationist, declared: "The moral tonic which the conservation movement has given to the entire country has perhaps been more effective for the general good of the American people than any one thing in our generation. . . ."[75] Such sentiments as these expressed the unifying factor among the new conservation enthusiasts, and tinged the conservation movement with the emotion of a religious crusade.

A wide difference in attitude separated Roosevelt, Pinchot, and Garfield from the new enthusiasts. The newer elements had little appreciation for rational and comprehensive planning, and the Roosevelt administrators, in turn, viewed with distrust the emotional fervor they aroused. Yet, Pinchot and his friends were forced to cultivate this sentiment to obtain the popular support which they desperately needed. The President who argued for national planning and efficiency in 1912 also "stood at Armageddon and battled for the Lord." As Pinchot discovered later, however, this marriage of convenience had its shortcomings. Although it was easy to whip up sentiment against a hostile Congress by portraying conservation as a moral issue, it was far more difficult to channel that same

[73] Scott, *Proceedings, 1910*, 270–277.
[74] Robert S. Yard, *Our Federal Lands* (New York, 1928), 239.
[75] Alfred L. Baker to Pinchot, Dec. 27, 1910 (GP #1818).

enthusiasm into support for specific laws. It was especially difficult to approach resource development in a rational manner when one's major political support now came from groups who looked upon the problem in moral rather than economic terms and preferred to reserve resources from economic use rather than to apply technology to their development.

Chapter VIII

Conflict over Conservation Policy

As the Roosevelt administration neared its final weeks, its resource leaders pondered the fate of the policies which they had fashioned. Would the President's chosen successor, William Howard Taft, pursue conservation affairs as vigorously as had Roosevelt? Many feared that he would not. In public speeches Pinchot and his friends assured their audiences that Taft approved of his predecessor's views on resource development and would continue established programs.[1] But privately they were not so certain.[2] Taft himself disagreed with much that Roosevelt had done, and in the field of water power policy for navigable streams Pinchot and Garfield had experienced this opposition firsthand. Taft revealed his doubts in a speech early in December 1908 when he cautioned, "There is one difficulty about the conservation of natural resources. It is that the imagination of those who are pressing it may outrun the practical facts."[3] From that time until March 4, 1909, the Roosevelt leaders feverishly formulated plans and issued directives which would render their policies invulnerable from the attacks of a hostile

[1] J. N. Teal, chairman, Oregon Conservation Commission in "Proceedings of the Joint Conservation Conference," in 60th Congress, 2nd Session, *Sen. Doc. 676, Report of the National Conservation Commission*, 25–27. Even Pinchot wrote Walter Fisher that "Taft is going to be with us in this work." See Pinchot to Walter L. Fisher, Feb. 23, 1909 (GP #120).

[2] Pinchot, *Breaking New Ground*, 377–379. Pinchot's and Garfield's fears are amply shown by the many executive acts taken during the first two months of 1909 to entrench the Roosevelt policies beyond any possibility of a change.

[3] "Proceedings of the Joint Conservation Conference," 124.

administration. And well they might, for Taft did modify much of the Roosevelt conservation program and, if he had had his way, would have altered it even more.

Roosevelt and Taft

Taft quarreled little with long-accepted measures, but he doubted the legality of many more recent executive acts within the spirit, though not the letter, of the law.[4] He disagreed especially with new hydroelectric, oil, and coal land policies which Pinchot and Garfield, interpreting their powers broadly, had devised for the public lands. While Taft's cautious and judicial predisposition led him in one direction, the attitude of Roosevelt men, built up through long years of administrative experience, led them in another. Smoldering for many months during the later years of Roosevelt's term of office and the first few months of Taft's administration, these disagreements flared into the open in the Pinchot-Ballinger controversy late in 1909. Though embellished and twisted by ingredients of political insurgency and muckraking, that conflict originally involved differences over specific features of the conservation program. On these particular policies Taft and his advisers radically disagreed with the Roosevelt administrators.

The two presidents differed primarily over the meaning of crucial resource laws, and over the role of the executive branch of the government in interpreting those laws. Relying upon such measures as the Forest Management Act of 1897, the Right-of-Way Act of 1901, and the General Dams Act of 1906, Pinchot and Garfield had expanded executive authority over the public lands and navigable streams. Taft, however, felt that these policies strained the intent of Congress. The new president, in fact, interpreted those measures even more narrowly than did the Supreme Court, which later upheld, as a legitimate exercise of the right to manage federal property, such innovations as the grazing fee and the water

[4] For example, Taft disapproved of Roosevelt's tactics in 1907 when he created over 50,000,000 acres of new forest reserves in the face of a congressional prohibition against such an act. See Henry F. Pringle, *The Life and Times of William Howard Taft* (2v, New York, 1939), I, 474–477.

power permit. Taft also argued that numerous executive commissions which Roosevelt had appointed had no legal standing. Each of these had provided the former president with vital information for policy recommendations, but Congress bitterly criticized them as an infringement on its prerogatives and Taft agreed. The new president discontinued the National Conservation Commission and dropped plans for a World Conservation Congress.

While Roosevelt desired a broad interpretation of law to enable the president to act effectively, Taft refused to exercise executive discretion unless it was specifically authorized by law. When a difference in interpretation of a statute arose, he preferred a limited view, no matter what the consequences for resource development. Roosevelt emphasized practical results, while Taft hesitated to increase executive discretion beyond that which Congress had specifically granted, and favored inaction if the meaning were ambiguous. Roosevelt's view enabled the federal government to attack problems which otherwise might not be solved, while Taft's attitude rendered the country vulnerable to those who sought to take advantage of a quiet and unassertive government. Roosevelt, impressed with the practical aspects of national problems, attempted to adjust political institutions to the technical requirements of an increasingly complex society. Taft did not fully visualize the need for that adjustment and often rejected its implications.

Taft's Conservation Advisers

The new president soon revealed that he would go his own way in resource affairs when he replaced Roosevelt administrators with others closer to his own point of view. This departure began late in December 1908, when Taft suddenly ceased to consult with leaders of the outgoing administration about his future official family, and rumors arose that he would not retain former cabinet members.[5] Although Taft earlier had informed Roosevelt that he would keep Garfield in the Interior post, on January 25, 1909 he notified the Secretary that he could not remain. Garfield, of course, had played

[5] This break between the Roosevelt and Taft groups can be traced in the diaries of James R. Garfield, beginning with the entry of December 22, 1908.

a key role in shaping the new public land policy with which Taft disagreed.[6] An even more crucial figure, the Assistant Attorney-General for the Department of the Interior, George Woodruff, received a "promotion" to the post of federal judge in Hawaii. At the same time, Department of Agriculture attorneys who had provided the legal basis for administrative water power policy, were transferred to the Solicitor's Office.[7]

These dismissals indicated to the Roosevelt group that Taft intended drastically to alter resource policy; his appointments to the cabinet confirmed their suspicions. The new Attorney-General, George Wickersham, had served as legal counsel for the Long Sault Development Company, a subsidiary of the Aluminum Company of America, which had sought and failed to obtain a perpetual and unlimited franchise for a large water power project in the Niagara River. In March 1908, at a conference with the president of the Long Sault Company, Roosevelt had first announced that he would veto future bills which did not provide for a rental and a limited franchise. Conservationists looked upon the Long Sault project as a crucial case, and upon anyone connected with it as suspect. Moreover, the Attorney-General substantiated their fears when in 1909 he announced his view that federal water power rentals were unconstitutional and that he had come to this conclusion as attorney for the power company.[8]

Conservationists received with greater alarm the appointment of Richard A. Ballinger to succeed Garfield as Secretary of the Interior. The Roosevelt group knew full well his views on resource matters, for, as Commissioner of the General Land Office from March 1907

[6] Pinchot argued that Garfield's dismissal was a result of pressure upon the Taft group by Alaskan coal claimants whom Garfield had forestalled. See Pinchot, *Breaking New Ground*, 405. It seems far more likely that Garfield's unusual role in Roosevelt's attempt to apply his water power policy to the navigable streams and in the administration's executive actions concerning public land questions aroused Taft's enmity.

[7] Wells, "Personal History" (GP #1671).

[8] *Ibid.* See also memorandum by Wells on St. Lawrence River Power (GP #1666) and memorandum on the Aluminum Company of America in the files of the National Conservation Association, 1911–12 (GP #1820).

to March 1908, he had disagreed violently with and had constantly criticized the administration's policies.[9] He had argued that the Bureau of Reclamation should release more lands and reservoir sites for private reclamation; he had proposed that Congress repeal the Right-of-Way Act of 1901, the legislative basis for administrative leasing of public land water power sites, and grant fee simple titles instead; he had objected to Roosevelt's coal land leasing measure; and, disputing the Pinchot-Garfield theory that unappropriated Western lands were fit only for grazing, he had opposed the grazing leasing plan.[10] The Commissioner had also objected to creation of the Chugach National Forest in Alaska, inclusion of grazing and untimbered land in the national forests in general, and administrative arrangements whereby the Forest Service played a role in deciding land claims in the national forests.[11] Since Pinchot, Garfield, and their friends had dominated resource matters under Roosevelt, Ballinger had found slight sympathy for his views. After eight months of conflict with the administration, early in November 1907 he decided to resign.[12] As Secretary, and under a president who had turned to other advisers, he would now be able to implement his policies.

Taft's new informal advisers created equal displeasure among the Roosevelt leaders. The new president, for example, frequently sought the counsel of his brother, Henry Taft, a member of

[9] See, for example, series of telegrams, James B. Adams to Pinchot, May 22, 1907; Adams to Pinchot, May 29, 1907; Adams to Pinchot, May 30, 1907; Pinchot to Adams, May 31, 1907; Wells to Pinchot, May 31, 1907 (all in RG #95, General Correspondence of Office of the Chief).

[10] Ballinger Report on Mondell Water Power Bill, Jan. 6, 1908, RG #49, Bill Book #29; Ballinger Reports on various coal bills, Jan. 22, 1908, Jan. 29, 1908, RG#49, Bill Book #29, pp. 395, 397, 399; Ballinger Reports on various leasing bills, Jan. 7, 1908, Feb. 26, 1908, in RG #49, Bill Book #29 and #30.

[11] Commissioner to Secretary of the Interior, April 24, 1907, May 10, 1907 (RG #49, Press Copies of Misc. Letters sent by Division "L"); Wells, "Personal History" (GP #1671).

[12] Ballinger informed Garfield on Nov. 2, 1907 of his intention to resign; see James R. Garfield diaries, entry of Nov. 2, 1907. One report indicated that a legal decision made by George Woodruff was the occasion for the resignation. *Collier's Weekly* (Aug. 28, 1909), 43, 7.

Wickersham's law firm, who had also served as legal counsel for the Long Sault Development Company.[13] It was Henry Taft who persuaded Ballinger to take the post of Secretary of the Interior.[14] The President, moreover, relied heavily upon the Corps of Engineers for guidance in resource affairs. Its members, in his view, were more levelheaded, less prone to follow popular clamor, and more cautious concerning their freedom of action within the constitution. In 1910, at Taft's insistence, Congress selected the Corps to undertake an investigation of irrigation projects.[15] In July of that year, the retired Chief of the Corps, Brigadier-General William L. Marshall, became consulting engineer to the Secretary of the Interior, and replaced Bureau of Reclamation officials as the President's personal representative to sessions of the National Irrigation Congress.[16]

These changes in personnel presaged changes in policy. Chosen partly because of their disagreement with the Roosevelt viewpoint, most of Taft's new advisers looked upon their appointments as an opportunity to redirect administrative action. In this they succeeded sufficiently to arouse a storm of protest from the Roosevelt leaders and to initiate the celebrated Pinchot-Ballinger affair. Historians have depicted this controversy as a struggle between personalities. Some describe Ballinger as a dishonest public official under fire from those who insisted on the highest standards of public morality. Others accuse Pinchot of unwarranted attacks on the Secretary of the Interior to vent his own personal feelings. Neither of these views correctly describes the origin of the heated controversy of 1909–10. For those differences concerned substantial disagreements over resource policy. The crux of the issue lay in the fact that former officials resisted changes which newer officials desired to institute.

Policy Changes: Irrigation

The new Secretary of the Interior, Richard A. Ballinger, hoped

[13] Wells, "Personal History" (GP #1671).

[14] John Hays Hammond, *The Autobiography of John Hays Hammond* (New York, 1935), 543.

[15] *Irrigation Age* (Aug. 1910), 25, 497.

[16] R. W. Young to Newlands, Sept. 23, 1912 (FGN).

to carry out a number of innovations in irrigation policy which from his experience as Commissioner of the General Land Office he felt to be desirable.[17] Realizing that Newell and other officials would resist his changes, he attempted to alter the personnel of the Bureau of Reclamation by offering the positions of director, chief statistician, and law officer to a number of prospects. But in this he failed, for those whom he approached declined the invitation. A city engineer in the state of Washington turned down the position of director on the grounds that nine-tenths of the Bureau's engineers would leave if Newell were fired. When Ballinger offered the post of chief statistician to the secretary of Senator Guggenheim of Colorado, pressure from "influential sources" forced him to withdraw the proposal.[18] Although these efforts were in vain, the new Secretary refused to seek advice from Newell and without the Director's knowledge, sent a personal representative to investigate irrigation projects in the West.[19]

Ballinger also attacked reclamation policies and reclamation officials to discredit them. He criticized Newell openly for beginning projects when money to complete them was not yet in the Reclamation Fund. He questioned the Director's engineering and administrative ability.[20] These attacks on Newell, persistent, yet often veiled, gave rise to spirited defense of him by many journals. Western papers, for example, demanded that the Secretary come out in the open and make himself clear. The *Denver News*, usually hostile to the Roosevelt resource program, admonished the Secretary, "It may be that Mr. Ballinger has a better way of doing things than Mr. Garfield had. If so, won't Mr. Ballinger please rise and explain?"[21] The *Engineering News* magazine, on the other hand,

[17] *Washington Herald*, Aug. 17, 1909 (FN #18, Clippings, v. 7).

[18] *Grit*, Aug. 15, 1909 (FN #18, Clippings, v. 7).

[19] *Washington Herald*, Aug. 19, 1909; *Spokane Press*, Aug. 9, 1909 (both in FN #18, Clippings, v. 7).

[20] *Leavenworth* (Kansas) *Post*, Aug. 12, 1909 (FN #18 Clippings, v. 7); *Colorado Springs Gazette*, Aug. 15, 1909 (FN #18 Clippings, v. 7). Taft joined in the criticisms; see *Irrigation Age* (Oct. 1909), 24, 465.

[21] *Spokane Press*, Aug. 9, 1909; *Denver News*, Aug. 11, 1909 (both in FN #18, Clippings, v. 7).

published a lengthy defense of Newell's ability, the accomplishments of the Bureau of Reclamation, and the public spirit of its employees who frequently turned down job offers from private firms at higher salaries because of their desire to contribute to the success of the federal program.[22]

The new Secretary achieved greater progress in modifying Bureau policy through legal decisions and administrative orders. In each case he argued that Newell had established practices not authorized by the Reclamation Act. Although some incidents substantiate this argument, the root of Ballinger's changes ran deeper: he wished to curtail federal reclamation work in favor of private irrigation development. As Secretary he pursued the policy of releasing for private ownership land and reservoir sites withdrawn for public projects. At the same time, although many urged him to do so, he would not press Congress to pass new laws to legalize practices which, he argued, Newell had illegally instituted.

Several examples illustrate the issues involved. Many irrigation projects had carried on their rolls an "irrigation farmer" to teach the newly arrived settlers how and where to run their ditches. For this advice from private sources the settler would normally have had to pay $20 to $30. The Bureau justified this practice by arguing that the law obligated it to bring water onto the land as well as to build dams and construct ditches. But Ballinger, declaring that he found "no warrant in law" for the irrigation farmer, abolished the position.[23] The Bureau had also employed "ditch riders," to inspect irrigation ditches regularly and to turn water from the main channels into laterals at regular intervals. Such employees the Bureau had found it impossible to hire for only a part of the year, and had used them elsewhere during the winter. Although the new Secretary approved the use of ditch riders, he insisted that they be employed only during the growing season when they performed their task of supervising the flow of water. Disregarding the practical problem, Ballinger merely argued that the Bureau could not legally employ them in winter.[24]

[22] *Engineering News* (Jan. 13, 1910), 63, 46–48.

[23] *Dayton* (Ohio) *Herald*, Oct. 4, 1909 (FN #18, Clippings, v. 8). [24] *Ibid.*

The most controversial reclamation issue involved the use of irrigation scrip to speed the program. Settlers who took up lands under projects in process of construction frequently could find no means of livelihood. At the same time, due to the lack of funds, the work progressed slowly. To solve both problems, Newell and Garfield devised a scheme whereby the Bureau hired the settlers and paid them in certificates which they could later return to the Bureau as part of their per-acre charge for the project. When first proposed, irrigation scrip had aroused considerable controversy. Although Woodruff had upheld its legality, others in the Department of the Interior had not. But the Bureau, with Garfield's approval, had instituted the scheme. Ballinger disapproved it and set out to abolish it. When the question was referred to Wickersham, the Attorney-General, he declared it illegal, arguing that the Secretary of the Interior could not let contracts exceeding in value the amount available in the Reclamation Fund.[25]

Policy Changes: Cooperation between Interior and Forestry

Until August 1909 the press concentrated primarily on disagreements between Ballinger and Newell. Throughout the summer, however, relations between Pinchot and Ballinger became increasingly strained, and from that time on the controversy between the Forest Service and the Department of the Interior occupied the limelight.[26]

Many national forest policies depended on close cooperation between the Interior and Agriculture Departments. For example, the Transfer Act of 1905 provided that the General Land Office would retain jurisdiction over all questions pertaining to land titles within the national forests. This cooperation, cordial under Garfield, broke

[25] Stahl, "The Ballinger-Pinchot Controversy," 88–89; *Portland Oregonian*, Sept. 19, 1909 (FN #18, Clippings, v. 7); *Newark* (N.J.) *News*, Aug. 19, 1909 (FN #18, Clippings, v. 7); *Boston Transcript*, Sept. 10, 1909 (FN #18, Clippings, v. 7).

[26] Pinchot complained to Elbert F. Baldwin, associate editor of the *Outlook*, that he could not work with Ballinger as he had with Garfield. See Pringle, *William H. Taft*, I, 479.

down when Ballinger became Secretary. When Pinchot tried to continue it, Ballinger complained that the Chief Forester sought to meddle in the affairs of his Department.[27] Under Roosevelt, Pinchot had frequently conferred with Interior Department subordinates without going through regular channels. When he attempted to maintain the practice under Taft he met stiff opposition from Ballinger and incurred the displeasure of the new president.[28]

Interdepartmental relations entailed far more than mere administrative channels. They involved different concepts of responsibility in resource management which led to mutual rivalry and suspicion. For almost a decade, the Forest Service, the Bureau of Reclamation, and the Geological Survey had argued that the federal government should possess large powers for constructive resource development. Choosing men through the civil service, these bureaus emphasized technical ability in their personnel. In their legal interpretations they encouraged broad construction of laws and expanded federal functions. In sharp contrast, the law clerks of the General Land Office were preoccupied almost solely with land title problems, and remained aloof from the new ideas in resource management. Citing its political appointees and its legalistic approach to land questions, officials of the newer departments openly criticized the General Land Office. The latter, in turn, rankled under these barbs. Disappointed when their spokesman, Commissioner Ballinger, left the service, Land Office personnel hoped that Taft would appoint him to the cabinet. Ballinger's successor as Commissioner, Fred Dennett, wrote him in November, 1908: "I am in hopes that the administration will not be able to get along without you in a cabinet position.... It certainly would be a source of everlasting joy to me to have you where I could reach you and consult with

[27] The Bureau of Reclamation, Ballinger complained, was more devoted to the policies and program of the Department of Agriculture than to those of the Department of the Interior. *Washington Times*, Nov. 18, 1909 (FN #18, Clippings, v. 8).

[28] See Taft to Helen Taft, Oct. 3, 1909 (WHT, Presidential Series 3, folder 1).

you."[29] When these hopes materialized, Land Office officials viewed the new turn of events as their opportunity to retaliate for a decade of criticism they thought unjust. Formerly they had not been able to resist Pinchot's and Newell's influence; but under a sympathetic Secretary they could now put these upstarts in their place.[30]

In the summer of 1908 Interior Department officials set out to change a number of cooperative arrangements between Interior and Forestry which Garfield and Pinchot had inaugurated.[31] The most controversial of these concerned the use of Forest Service personnel to supervise fifteen million acres of Indian Reservation forest land. Since the Indian Office, in the Department of the Interior, had no trained foresters, it frequently consulted the Forest Service on technical questions. This practice led to a formal cooperative agreement between the two agencies in January 1908, in which the Forest Service agreed to supervise timber cutting and fire protection activities on the reservations, and the Indian Office agreed to pay the salaries and other expenses involved. During the summer of 1908 the Forest Service carried out field work according to this agreement, but it was not pleased with the handling of the related routine work in the Indian Office. To solve this problem, Pinchot suggested that an experienced clerk from the Forest Service be transferred to the Indian Office.[32] But when the Comptroller-General ruled that the proposal was illegal, the Chief Forester went no further.[33]

A few months later Ballinger called into question the entire cooperative arrangement. When Indian officials considered the appointments for fire control work for the summer of 1909, the Secretary insisted that the men be carried on the rolls of the Interior

[29] Dennett to Ballinger, Nov. 28, 1908 (RG #48, file 1–108, Congressional Investigation of Interior Department).

[30] M. D. McEniry to Fred Dennett, Sept. 4, 1909; McEniry to Dennett, Sept. 10, 1909; McEniry to H. H. Schwartz, Aug. 31, 1909; McEniry to Dennett, Aug. 1909 (all in GP #1827).

[31] This problem can be followed in *Hearings, Interior and Forestry*, v. 2, 88–97; v. 4, 1153–1154, 1186–1206; v. 6, 3042–3045, 3123-3130.

[32] See Pinchot MSS on Indian lands in GP #2139.

[33] Robert Tracewell to Secretary of Interior, Sept. 3, 1908 (GP #1672).

Department, subject to Indian Office control. A few days later he notified the Indian Commissioner and Secretary of Agriculture that the Comptroller-General's decision in the clerk case had rendered the entire agreement illegal, and that the Indian Office would henceforth establish a separate forest administration for the Indian Reservations. Use of the Comptroller-General's decision to end the cooperative agreement surprised not only Pinchot and the Forest Service but also two staunch Taft supporters in the Senate, Elihu Root and George Sutherland, who argued that the decision had pertained solely to the clerk and not to the entire cooperative agreement.[34]

Divided jurisdiction over Indian timberlands had prompted Pinchot to devise a plan to transform them into national forests. To effect this change with Indian reservations which Congress had created required congressional authorization. But the president could himself transfer reservations originated by executive order. After working out the details with the Office of Indian Affairs, Pinchot and his law officers prepared executive orders which simultaneously restored these forest lands to the public domain and reserved them as national forests. Two days before the end of his administration, Roosevelt issued proclamations to this effect. From that day, March 2, 1909, by the terms of these orders, the lands were subject to the jurisdiction of the Forest Service. Yet, Ballinger refused to yield authority over them, and the Forest Service did not receive charge of their management. The new Assistant Attorney-General, while not attacking the legality of the withdrawals, indicated that the Indian Office, rather than the Forest Service, should sell the standing timber. Congress reinforced this opinion with a general act on June 25, 1910 which authorized the Indian Office to sell mature live timber on its reservations. Not until 1912 did Taft issue orders to exclude the seven reservations from national forests and restore them to their former status as Indian reservations.[35]

[34] Indian Commissioner Valentine wrote that he never had been informed that the decision of Sept. 3, 1908 "bore in any way on the general legality of the cooperative agreement." Valentine to Pinchot, Oct. 8, 1909 (GP #1672).

[35] The best description of this problem is in Wells, "Personal History" (GP #1671).

Still a third controversy arose over land claims and titles within the national forests. The Forest Transfer Act of May 1, 1905 had provided that national forest land title questions remain under the jurisdiction of the General Land Office. Yet, the Forest Service felt it necessary to control these matters itself. For example, since the Act of 1897 had opened national forests to mineral entry at all times, one could readily use a mining claim to gain control of valuable forest, range, or other lands within the forest boundaries. The General Land Office had no jurisdiction to forestall such false entry. Mineral claims were registered only in local county offices, and did not appear on the records of the federal government until the prospector applied for a patent. The Land Office argued that it did not have authority to examine the validity of a mineral claim merely recorded in a county court; and it did not even investigate the validity of a patent itself on the ground unless it was contested. The Forest Service hoped to forestall illegal entries and patents from the start by examining them in detail. Pinchot, therefore, worked out an arrangement whereby the Department of the Interior decided all questions of title, and the Forest Service investigated and determined all questions of fact with respect to claims and patents within the national forests. With Roosevelt's approval, both Secretary of Agriculture Wilson and Secretary of the Interior Hitchcock signed the agreement in May 1905.[36]

As Commissioner of the General Land Office, Ballinger had argued that this arrangement was illegal and interfered with Land Office prerogatives. The Comptroller-General sustained this view, by ruling that, since Congress had already appropriated funds to the General Land Office to investigate land claims, the Forest Service could not legally duplicate this work. An appeal to the Attorney-General produced a working agreement whereby both departments separately collected information and then exchanged each other's records at will. To render this arrangement more effective,

[36] *Ibid. Decisions of the Department of the Interior and General Land Office.* v. 33, p. 609 (Washington 1905). See correspondence on this question, Roosevelt to Secretary of Agriculture, May 17, 1905 (TR); Roosevelt to Hitchcock May 17, 1905 (TR).

the Forest Service successfully attached to the 1908 Agricultural Appropriations bill a rider providing that the Secretary of the Interior could request information from the Forest Service concerning land claims. Secretary Garfield frequently made such a request, and on March 2, 1909, drew up a memorandum formalizing the procedure.[37]

Secretary Ballinger honored this agreement, save for one point, which to the Forest Service was extremely vital. In most cases, the two departments had agreed, the Forest Service would furnish information to the Land Office, which would then prosecute illegal entrymen. But in cases involving rights-of-way across national forests, the Forest Service could institute proceedings on its own initiative so that it could control actions against violators of its water power permits. Some of the hydroelectric corporations operating plants in the national forests prior to the new permit system had refused to accept Forest Service regulations. After the courts had decided one test case in its favor, the Forest Service instituted proceedings against the remainder. Ballinger, however, refused to surrender control over these legal proceedings, and demanded that the Forest Service submit the facts in each instance to the General Land Office, which, in turn, would prosecute the case. Most of these actions remained dormant in the courts during Ballinger's administration and were taken up only later by his successor.[38]

Policy Changes: Water Power

The Taft administration's new water power policy was its most far-reaching innovation.[39] The President and his subordinates disagreed radically with Roosevelt's approach toward hydroelectric

[37] Wells, "Personal History" (GP #1671); Garfield to "Sir," Mar. 2, 1909 (GP #1672).

[38] *Ibid.*: Ballinger to Wilson, Mar. 17, 1909; Apr. 14, 1909 (GP #1672); Wells, "Personal History" (GP #1671).

[39] The water power problem for the public lands can be traced in *Hearings, Interior and Forestry*, v. 2, 80–87, 540–608; v. 4, 1154–1186, 1492–1505; v. 5, 1637–1656, 1670–1674, 1678–1765; v. 6, 3035–3040, 3130-3137, 3149-3157.

power on both the public lands and navigable streams. They set out especially to reverse Garfield's major innovation of applying to the public lands the permit system which Pinchot had established for the national forests. As did Pinchot, Garfield interpreted the Right-of-Way Act of 1901 as a congressional mandate that lands chiefly valuable for water power development should be permanently reserved for that use.[40] To carry out this objective Garfield began to devise a permit system and to withdraw public land water power sites from all forms of entry except under this act.

Taft and Ballinger argued that the Right-of-Way Act did not grant authority to establish the permit system, but simply provided a temporary license pending the transfer of title to private parties. Ballinger directed the General Land Office to warn all licensees that their privilege would terminate as soon as Congress enacted new laws. A permit issued in 1910 to the Minnesota Canal and Power Company, for example, stipulated only one restriction—that the permittee promise prompt and full conformity to present and future laws and regulations governing the lands involved. Although the Secretary could not similarly reverse the new system in the national forests, he did argue, with Wickersham, that the Forest Service suits against permit violators rested on shaky legal ground; therefore he postponed those proceedings.[41]

In the President's view, the federal government, as riparian owner of Western lands, could constitutionally control hydroelectric installations on the public domain, including a limited franchise, rentals, and forfeiture in cases of violation. Speaking before the National Irrigation Congress in Spokane in September 1909, Taft urged Congress to enact these provisions into law.[42] But Ballinger protested against this policy. In his second annual report as Secretary he wrote: "In the various public land states and territories containing water-power resources . . . it must be realized that any radical or burdensome restriction imposed by the federal government upon the resource will operate as a servitude on the public lands

[40] See Wells to Elbert F. Baldwin, Apr. 28, 1910 (GP #1670).

[41] *Ibid.*; Pinchot to S. A. Kean, Aug. (nd), 1910 (GP #1819).

[42] *Proceedings, National Irrigation Congress, 1909,* 523–525.

and discourage their development and use."[43] By increasing the cost of hydroelectric power, the Secretary argued, the national forest permit system thwarted Western development, and extension of that system to the public domain would retard progress even further. Ballinger would solve the problem by ceding the water power sites to the states, without restrictions as to rentals. Senator Reed Smoot of Utah introduced a bill for this purpose in the fall of 1909 and Ballinger endorsed it the following spring. In May, after Congress had not acted on his earlier proposal, Taft publicly approved the Smoot bill; when it failed to pass he reverted to his former request for full federal regulation.

Ballinger and Taft also questioned the legality of Garfield's power site withdrawals from all entry except under the Right-of-Way Act of 1901. Over the protests of the chief officers of the Bureau of Reclamation, the Secretary restored these lands to all forms of entry. But, after Pinchot protested to Taft against this action and Taft prodded Ballinger, the Secretary re-withdrew the sites. These were exactly the same areas which Garfield had withdrawn; yet from their superficial description it appeared that Ballinger had reserved the same sites by setting aside a smaller area. The Garfield order had described the land by sections, and in a general statement had excluded from its provision the many private holdings within these boundaries. Ballinger, however, specifically subtracted the private acreage from the total and boasted that with a smaller area he had accomplished the same end as had Garfield, and whereas his predecessor had included private holdings in his withdrawals, his own were confined to the public lands. Neither statement was correct.[44]

Ballinger's action differed from Garfield's in an even more significant manner, for the new Secretary withdrew the sites from *all* forms of entry rather than from entry *other than* under the Right-of-Way Act.[45] This, in effect, halted any progress toward leasing

[43] *Ann. Rept. Secty. Int., 1910,* 19.

[44] See memorandum by A. P. Davis, Chief Engineer of the Bureau of Reclamation in GP #120.

[45] For the wording of Ballinger's withdrawals and its contrast with Garfield's, see *Hearings, Interior and Forestry,* v. 2, pp. 569, 577–578. See also Wells to Arthur W. Page, Aug. 15, 1910 (GP #1670).

public domain water power. Ballinger, moreover, religiously extended his order to every possible public land power site. Lands excluded from national forests were immediately withdrawn from entry to determine if they could be developed for power.[46] He suspended all private irrigation right-of-way entries under the Act of 1891 and rejected those which contained power sites.[47] And he unsuccessfully tried to persuade the Forest Service to withdraw its potential power areas from all forms of entry including under the Right-of-Way Act of 1901. The Forest Service "refused to take this action for [the] reason [that] it would block development."[48] The Secretary's action, therefore, halted hydroelectric development on the public domain while it went forward steadily in the national forests. Garfield had sought to reserve areas for a specific use, but Ballinger intended to prevent any use and thereby force a decision in Congress on the water power question. Roosevelt administrators complained bitterly that the Secretary was thwarting Western development, but the public, not understanding the issue, failed to catch the significance of Ballinger's action and accused Pinchot and Garfield of withholding lands from present use for future development.

Though forced to carry out the re-withdrawal, both Taft and Ballinger doubted its legality and asked Congress to enact a law permitting such action.[49] The limited nature of this request revealed that the old differences persisted. Ballinger, in his proposed measure, asked that the withdrawals be only temporary, pending congressional action on the water power question. Roosevelt leaders retorted that Congress had already acted in the Right-of-Way Act of 1901 and that Ballinger was merely maneuvering to defeat a public regulation policy under that Act.[50] The Withdrawal Act, which Congress approved on June 25, 1910, authorized the president to

[46] Roosevelt to Pinchot, Jan. 17, 1911 (TR).

[47] Wells to Page, Aug. 17, 1910 (GP #1670); Pinchot to Kean, Aug. (nd), 1910 (GP #1819).

[48] Wells, "Personal History" (GP #1671).

[49] See statement of Taft in 61st Congress, 2nd Session, *House Doc.* 535, 4–5.

[50] Wells to E. A. Sherman, Jan. 23, 1911 (GP #1665); Wells to Page, Aug. 15, 1910 (GP #1670); Wells, "Personal History" (GP #1671).

withdraw land *temporarily* for any public purpose. Pinchot, however, succeeded in securing an amendment which provided that "such withdrawals or reservations shall remain in force until revoked by him [the president] or by an Act of Congress." Conservationists hoped that this wording would permit permanent withdrawals and, based upon them, an administrative water power policy for the public domain. Attorney-General Wickersham shattered these plans when, on September 23, 1910, he ruled that Congress intended the Act "to designate and set apart for future action by Congress the areas of conspicuous value as water power sites. . . ." The administration still believed that the Right-of-Way Act of 1901 granted it no power to lease.[51]

Taft took an even stronger stand on hydroelectric development in navigable streams. He had always believed that the federal government constitutionally could charge a rental for power produced at dams constructed by the federal government to improve navigation, but could not do so at privately built ones.[52] "The mere duty," he wrote, "to apply to the Government for a permit [to build a private dam], which means only a declaration by the Government that it will not interfere with the navigability of the stream, does not give to the Government the power to impose conditions for its use, except so far as those conditions are necessary to preserve navigability in the stream."[53] The commerce clause of the Constitution, so Taft argued, limited the power of the federal government to navigation regulation and to those aspects of water development directly affecting navigation.

He reversed Roosevelt's policy by signing numerous bills granting perpetual and unlimited franchises for the construction of dams in navigable streams. One of these, the James River measure, Roosevelt had vetoed in January 1909. Congress balked only when the Long Sault project, involving a potential development of 600,000 horsepower, appeared once more.[54] Yet, Taft continued to argue that he

[51] Wells to E. A. Sherman, Jan. 23, 1911 (GP #1665).

[52] 60th Congress, 2nd Session, *Congressional Record*, 2607.

[53] Taft to Pinchot, Oct. 12, 1909 (GP #120).

[54] *American Conservation* (Mar. 1911), 1, 2.

had no power to modify the situation, and Attorney-General Wickersham sustained this view. In an interpretation of the General Dams Act of 1906 the latter argued that the law authorized neither rentals nor conditions to control monopoly. Although that Act seemed to grant broad regulatory power, said Wickersham, a crucial clause had provided that regulation might include provisions for a lock at the licensee's expense, and therefore had narrowed regulation to the field of navigation.[55]

Wickersham's decision prompted Congress to pass a new General Dams Act, June 23, 1910, which contained only a slight concession from the administration.[56] If federal reservoirs and forests actually improved navigability and water supplies for the power corporations, Taft maintained, then a fee could be charged, but only to the extent of defraying the cost of federal public improvements. The new act provided for such a fee. House leaders regarded this provision as a "compromise" with Roosevelt and Pinchot. Yet its method of determining rentals proved unworkable.[57] Moreover, since the Army Corps of Engineers did not believe that headwater forests affected stream flow, that agency did not impose the charge provided by the law.[58]

The Pinchot-Ballinger Controversy

Roosevelt administrators reacted sharply to these changes in resource policy. They now faced the alternatives which Ballinger had encountered in 1907—either to acquiesce quietly or to leave the government service. Some, such as the Commissioner of Indian Affairs, Francis E. Leupp, resigned. But most, especially those in the Reclamation Bureau, remained and attempted to stem the tide. Although publicly they denied that a rift had arisen between

[55] 27 *Opinions of the Attorney General*, 462; see memorandum, "Wickersham's Connection with Water Power," in GP #1665.

[56] Kerwin, *Water-Power Legislation*, 128.

[57] William C. Adamson to Lindley M. Garrison, Secretary of War, Oct. 4, 1913 (RG #77, #49678/105); Kerwin, *Water-Power Legislation*, 129; Chief of Engineers to Secretary of War, May 14, 1914 (RG #77, #49678/125).

[58] *Portland Oregonian*, Oct. 5, 1909 (FN #18, Clippings, v. 8).

themselves and the Secretary, privately, after vigorous protest, they accepted many changes.[59]

The new Secretary of the Interior could not as easily control Chief Forester Pinchot, an official of another department, whose activities remained as freewheeling as they had been in the previous administration. Pinchot assumed the leadership of the attack on the new Taft policies. When Ballinger prevented the Forester from contacting Interior Department officials directly, Pinchot took his case to Taft. In April his intervention prompted the President to direct his Secretary of the Interior to re-withdraw the restored water power sites. Three months later he warned Taft that Ballinger's criticisms of the Bureau of Reclamation had discouraged and demoralized officials of that agency. To persuade the Interior Department to pursue Garfield's public lands water power policy, Pinchot worked out with Attorney-General Wickersham a proposed form of power contract; however, when he submitted it to Ballinger, the Secretary turned it down.[60]

Pinchot tried desperately to change Taft's point of view, but with little success. The Forester argued that Ballinger's actions indicated a lack of sympathy with the basic objectives of resource policy which Roosevelt had developed. Taft replied that, while the administration wished to implement a broad and effective policy, it also intended to keep strictly within the law. In turn, Pinchot argued that many laws, unclear in their meaning, should be construed broadly to render them most effective. By late November 1909 Taft had lost patience with such argument. He implied to Pinchot that he wanted to hear no more of his criticism of Ballinger and that if the Forest Service would be quiet, everything would work out all right. At the same time the President instructed the Secretary of Agriculture to clamp down on Forest Service publicity.[61]

[59] See statement by Newell in *Yakima Republic*, Aug. 7, 1909 (FN #18, Clippings, v. 7); memorandum by A. P. Davis (GP #120).

[60] Pinchot, *Breaking New Ground*, 406–407, 409, 436–438. See also Pinchot MSS in GP #2139; *Washington Times*, Oct. 28, 1909 (FN #18, Clippings, v. 8).

[61] Pinchot, *Breaking New Ground*, 436–437; see especially Taft to Pinchot, Nov. 24, 1909, 438.

Taft's admonitions in the fall of 1909 sharpened Pinchot's dilemma and moved him further toward a break with the President. Pinchot's friends had urged him to resign early in the summer, but he had remained at his post, hoping to modify the administration's point of view.[62] At the same time, however, the Forester had prepared the way for an eventual break which would transfer the battleground from the inner circles of the administration to the general public. During the summer of 1909 he had organized the National Conservation Association to create widespread popular support for the Roosevelt conservation policies. In the fall he laid plans for a national magazine to increase this support once the battle with the administration came into the open.[63] Taft urged Pinchot not to resign. Some of his advisers had tried to persuade him to fire the Forester, but he hesitated for fear that Pinchot would invoke Roosevelt's name and prestige in opposition to him.[64]

During the late summer and early fall of 1909, Pinchot's criticism of the administration and the Interior Department became crystallized in the Alaskan coal land case.[65] A young agent for the Department of the Interior, Louis R. Glavis, felt that Ballinger had improperly helped several corporations to acquire larger holdings of Alaskan coal land than the law permitted. After protesting to the

[62] Henry Graves to Pinchot, Sept. 27, 1909 (GP #91).

[63] Pinchot to Charles W. Eliot, Aug. 5, 1909 (GP #120); Emerson Hough to Pinchot, Nov. 12, 1909; Pinchot to Hough, Nov. 16, 1909; Hough to Pinchot, Nov. 23, 1909 (all in GP #1809).

[64] Taft to Pinchot, Sept. 13, 1909 in Pinchot, *Breaking New Ground*, 430–431. George Mowry, *Theodore Roosevelt and the Progressive Movement* (Madison, 1947), 79.

[65] The most complete reviews of the Alaskan coal case are in Stahl, "Ballinger-Pinchot Controversy," and A. T. Mason, *Bureaucracy Convicts Itself: The Ballinger-Pinchot Controversy of 1910* (New York, 1941). See also Harold L. Ickes, *Not Guilty, An Official Inquiry into the Charges Made by Glavis and Pinchot Against Richard A. Ballinger, Secretary of the Interior, 1909–1911* (U. S. Dept. Int., Office of the Secretary, 1940), and Pringle, *William H. Taft*, I, ch. 26–27. The major body of evidence is in *Hearings, Interior and Forestry*, v. 2, pp. 4-80, 97-103, 106-539, 609-805; v. 2; v. 4, 771-1152, 1207-1212, 1238-1257, 1261-1281; v. 5, 1515-1547, 1674-1677, 2048-2410; v. 6, 2411-3035, 3077-3123.

Department in vain, Glavis took his case to Pinchot. Since some of the lands in dispute lay in a national forest, Pinchot took the initiative in persuading Taft to look over the land agent's charges. Taft was in no mood to investigate the matter thoroughly. Irritated by the needling tactics of Midwestern Insurgents who used conservation issues for ammunition, and convinced that the Forest Service inspired and supplied information for these attacks, he chose to rely upon the judgment of Wickersham and Ballinger. Following their advice, Taft cleared the Secretary of the Interior of all wrongdoing, accused Glavis of presenting a one-sided case, and fired him.

Early in November *Collier's* magazine published Glavis' charges against Ballinger. At once the issue became nation-wide and the focus of the entire controversy over resource policy. In fact, the Alaskan coal claims completely transformed the conflict within the administration from differences over resource development into a question of Ballinger's personal character. One suspects that the Roosevelt leaders, perhaps without deliberate intent, centered their fire on Ballinger's integrity because with that issue they could make their case clear. While the public could scarcely fathom the intricacies of administrative action under the Right-of-Way Act of 1901 or multiple-purpose river planning, it well understood the misconduct of a public official.

By late 1909 Pinchot had decided to "make the boss fire him."[66] Early in January 1910, in a letter to the Senate, Ballinger invited a full investigation of his department. Accusing the Forest Service of responsibility for attacks on him, the Secretary recommended that the investigation include that bureau as well. In a reply to these accusations, written to Senator Dolliver, Insurgent Republican from Iowa, and read on the Senate floor, Pinchot defended both the Service and the officials whom Ballinger had accused. Seizing upon this

[66] As early as September 24, 1909 Pinchot had warned Taft that "he might be forced to fire me." See Pinchot, *Breaking New Ground*, 434. In 1913 his close friend, Wallace D. Simmons, wrote Pinchot, with obvious reference to the Pinchot-Ballinger controversy, that he (Simmons) was in no position to "'make the boss fire me.'" See Simmons to Pinchot, Feb. 3, 1913. See also Mowry, *Theodore Roosevelt*, 80.

letter as an act of insubordination, Taft immediately fired the Chief Forester.

From January to April 1910 a Joint Committee of the Senate and the House investigated the activities of both the Department of the Interior and the Forest Service. Through forty-five days of testimony, cross-examination, and interparty wrangling, the Committee thoroughly aired the policies and viewpoints of both Pinchot and Ballinger on conservation issues. The investigation, however, did not produce a sober evaluation of resource policy. Although some, such as Senator Elihu Root of New York, a staunch defender of the Taft administration, hoped that the hearings would provide a basis for constructive conservation legislation, few of the participants concentrated their testimony on that problem. To Ballinger the occasion provided an opportunity to vindicate his personal integrity. Taft hoped that it would deflate the Chief Forester whom he regarded as "too much of a radical and a crank." The Insurgents, not fundamentally interested in the Roosevelt brand of conservation, desired to embarrass the standpat wing of the Republican party, and Taft himself, whom they accused of selling out to the "party bosses." And Pinchot concentrated on proving that Ballinger was dishonest and Taft inept in covering up for his Secretary. Few constructive conclusions arose from this trial of the entire Taft administration.[67]

The Joint Committee reports reflected the atmosphere of the hearings; their conclusions followed the battle lines drawn months before. The majority of regular Republicans vindicated the Secretary and the entire administration, while the two minority reports, one by the four Democrats and the other by the single Insurgent, censured them. On matters of specific resource policy the reports divided in the same manner, yet these conclusions fell by the wayside as the censure and defense of the administration assumed the spotlight. Taft felt that his Secretary had received a clean bill of health, while the opposition labeled the majority document a "whitewash." The enormous quantity of factual data uncovered merely provided ammunition for each side in the struggle, and especially

[67] These conclusions are based on *Hearings, Interior and Forestry*.

enabled the Insurgents to intensify their attack on Taft. To the disinterested student of public affairs the hearings were a gold mine of information about resource affairs, but amid the storm of partisan strife, such an individual, if he existed, could gain no hearing.[68]

The Legacy of the Controversy

For years to come as well as in 1910 the Pinchot-Ballinger controversy obscured rather than clarified conservation problems by reducing complex questions of resource management to simple matters of personal honesty. A lack of public morality did not generate the debate; its root lay in differences over administrative policy which emerged as early as 1907. Pinchot did little to clarify this confusion. For, as Taft wrote, with some accuracy, "His [Pinchot's] trouble is that no one opposes his methods without arousing in him a suspicion of that person's motives. . . ."[69] The Chief Forester insisted to the investigating committee that he had nothing personal against Ballinger; yet he actually looked upon the Secretary almost as a thief who wanted to steal the public lands and upon Taft as a co-conspirator.[70] Moreover, Pinchot permitted the entire controversy to become involved with the simple Insurgent tactics of discrediting the Taft administration. These tactics rallied a vast public support against Taft and Ballinger and temporarily unified conservation sentiment. But for years the consequences remained a millstone around the necks of those who wanted to think clearly and act concretely about specific conservation problems.

Yet the affair did result in a number of important policy changes. The Alaskan claims did not come to patent, and Taft hesitated to permit more public lands to fall into private hands. The strong political pressure which the controversy enabled the Chief Forester to bring to bear upon the administration, however, should not be overemphasized. Despite urgent appeals, neither Taft nor Ballinger would undertake the Roosevelt program for waterways, water

[68] *Ibid.*, vol. 1.

[69] Taft to Helen Taft, Oct. 3, 1909 (WHT, Presidential Series 3, folder 1).

[70] Roosevelt to Alvord Cooley, Nov. 23, 1910; Roosevelt to William Kent, Nov. 28, 1910 (both in TR).

power, or mineral lands. Moreover, it was Taft, rather than Pinchot, who prevented his Secretary of the Interior from going further in modifying former policies. Taft disagreed with Pinchot largely over methods of executing policy, while Ballinger disapproved of the substance of the Rooseveltian policy itself. On matters of policy, the Secretary later acknowledged, he had not been his own master; "In the preparation of legislation and its advocacy . . . I had to conform to the views of the administration, my personal views not being wholly in harmony with those of the President. . . ."[71] Although some have argued that pro-Pinchot sentiment forced Ballinger to resign in March 1911, the evidence indicates that he left voluntarily because Taft failed to give him free rein to implement his viewpoint. Scarcely a few months later, moreover, he could be found giving encouragement to those in the West who attacked the administration for not disposing of Western public lands to private individuals.[72]

The Pinchot-Ballinger controversy played a more important role in the selection of two new cabinet members. In order to heal the breach in the Republican party, and thereby to assure his renomination in 1912, Taft held out the olive branch to the Republican opposition. He appointed two men favorable to the Pinchot point of view—Walter L. Fisher as Secretary of the Interior, and Henry L. Stimson as Secretary of War. Active in conservation affairs, both Stimson and Fisher agreed with Pinchot on specific policies but had frowned upon his excessive zeal and his sponsorship of the public crusade of 1908–1910.[73] As Secretaries, they carried on with policies which Garfield and Pinchot had begun to develop in 1907–9. Fisher worked out a water power permit system for the public lands, and even considered administrative action toward mineral leasing. Stimson backed both the Newlands waterways bill and the Roosevelt program for hydroelectric development on navigable streams.

[71] Ballinger to J. Arthur Eddy, July 15, 1911 (WHT, Presidential Series #3, 1909–13, file #114).

[72] *Ibid*; Ballinger to Taft, July 24, 1911; Ballinger to Taft, July 28, 1911 (WHT, Presidential Series #3, 1909–13, file #114).

[73] Roosevelt to William Kent, Nov. 28, 1910 (TR).

Yet, their efforts for mineral and water power leasing did not bear fruit until Congress acted in 1920.

The conflict between Roosevelt and Taft policies also clarified departmental relations and halted steps to coordinate water and land resource administration. Through executive action and recommendations to Congress, Roosevelt had tried to overcome departmental resistance to a more comprehensive approach to conservation affairs. Yet this policy had been vulnerable, because interdepartmental cooperation had come largely through the personal influence of Gifford Pinchot with the President. For all practical purposes, a departmental subordinate, a bureau chief, was Roosevelt's Secretary of Agriculture and Interior. Taft considered this practice to be not administrative coordination, but "demoralization of discipline that follows the reposing of such power in the hands of a subordinate."[74] He spoke from personal experience, for in 1908 Roosevelt had forced him, as Secretary of War, to follow Pinchot's advice in recommendations to Congress on water power measures. Departmental officials not close to the Pinchot-Garfield group considered the Forester a meddling outsider. The Pinchot-Ballinger controversy completely destroyed this "coordination." Ending many of the interdepartmental executive practices of his predecessor, and failing to recommend that Congress establish closer coordination, Taft more firmly intrenched the Corps of Engineers and the Department of the Interior in their respective bailiwicks and thereby dealt a blow to more comprehensive resource management.

The simplicity and ambiguity of the crusade against the Taft administration unified, temporarily, many who actually disagreed with the Roosevelt policies. Insurgent leaders, for example, responded with delight to the former President's attacks on business, and derived political advantage from his campaign for river development, but they feared his zeal for the efficiency of bigness and for independent executive action, both of which his conservation policies often revealed. Never before had they moved in the inner circles of the Roosevelt conservationists. The fight against Ballinger also brought park and civic improvement organizations into close co-

[74] Taft to Helen Taft, Oct. 3, 1909 (WHT, Presidential Series #3, folder 1).

operation with the Roosevelt leaders. Yet, specific issues had revealed that both Pinchot and Roosevelt frequently favored commercial development when it conflicted with aesthetic interests. Once the fervor of struggle against a common opposition died down, the old differences between such groups reemerged with increasing intensity.

The controversy had even more long range consequences; from it stemmed conceptions about the conservation movement that were false, yet firmly rooted in the public mind. To argue, as many thereafter did, that conservation entailed a crusade for the "common people" against the "trusts," grossly oversimplified the case.[75] Some large corporations staunchly supported the Roosevelt program, while many "people," such as Western farmers, violently opposed it. The Pinchot-Ballinger controversy, moreover, established the permanent iniquity of the Department of the Interior as a co-conspirator in alienating the public lands, while the Forest Service remained the guardian of all that was good and wise in protecting resources in the public interest.[76] However, with time and new Secretaries, the Interior Department also became a public land management agency.

An equally false version of the episode was that the controversy involved issues of personal integrity, with Ballinger the villain and Pinchot the hero. Although the late Secretary of the Interior Harold L. Ickes reversed the moral roles, casting Pinchot as the frustrated official who sought revenge on an innocent public servant, his attempt to defend the Interior Department still obscured the essential differences over policy amid the conflict of personalities.[77] Still a fourth misconception arose from Pinchot's leadership in the attack on Ballinger, for in the minds of the general public it placed the Chief Forester head and shoulders above all in the conservation movement. Actually, others made contributions of vast significance. While this emphasis maximized problems of the public domain, it obscured water development in navigable streams and neglected the

[75] Examples of this over-simplification are Pinchot, *Breaking New Ground*, 504–510; Stahl, "Ballinger-Pinchot Controversy," 1.

[76] For some effects of this view see Peffer, *Public Domain*, 232–246.

[77] Ickes, *Not Guilty*.

role of the Corps of Engineers, Newell, Newlands, Maxwell, and Leighton in the conservation movement.[78]

[78] Pinchot's book, *Breaking New Ground,* illustrates this tendency. Although he includes pictures of many Forest Service officials, as well as McGee, he has no pictures of Newell, nor does he even mention Marshall O. Leighton. Another example of this tendency is a recent source book in American history, Thomas G. Manning and David M. Potter, *Government and the American Economy, 1870–Present* (New York, 1950), 161–187, in which the section on conservation concentrates primarily on forestry.

Chapter IX

Organized Conservation in Decline

Conservation, so its partisans believed, was a new way of looking at all public problems. The American people, once convinced of this, would demand with a single voice that its principles be adopted.[1] Yet, conservation in practice meant vastly different things to different people. The movement's unity, as exhibited by the intense emotional fervor between 1908 and 1910, proved to be false, a religious enthusiasm directed against certain federal policies and officials rather than a common support for agreed-upon positive measures. As concrete issues became clarified, diverse interests revealed this superficial unity and shattered the unified crusade into particularistic groups.

In fact, the broad outburst of the years of the Governors' Conference and the Pinchot-Ballinger controversy carried the seeds of the movement's disintegration. Resource organizations with more limited aims feared that the larger movement would engulf their objectives, and left conservation bodies soon after joining them. The waterway associations, for example, after supporting Pinchot in the winter of 1907–8, shortly went their own way. Similar groups became dissatisfied as the formal conservation associations limited their objectives to the specific issues which Pinchot thought to be most important. The movement's popular support receded rapidly; its leaders quarrelled among themselves. By the First World War

[1] Pinchot, *Breaking New Ground*, 504–510; Pinchot to Walter L. Fisher, Feb. 23, 1909 (GP #120).

conservation as an organized force had shriveled to a small group of men dominated by the influence, personality, and interests of its leading figure—the former Chief Forester, Gifford Pinchot.

The Expanding Movement

The term "conservation" proved to be highly elastic. After the summer of 1908 its meaning expanded to include almost every movement of the day, and a wide variety of reformers flocked into conservation organizations. Roosevelt initiated the process by popularizing the conservation of human health. With Professor Irving Fisher of Yale as its head, a "Committee of 100" had advocated a federal public health program. At the suggestion of the Committee, Roosevelt prepared plans for a new department of public health and included an investigation of national health conditions in the work of the National Conservation Commission.[2] In 1910 the National Conservation Congress organized a standing committee on "vital resources," with five men interested in public health as members, as well as units on forests, lands, waters, and minerals.[3] Two years later the Congress devoted its entire annual session to "the conservation of human life."[4]

This was only a beginning. At the first session of the National Conservation Congress in August 1909 delegates delivered speeches on the conservation of peace and friendship among nations, the conservation of the morals of youth, the conservation of children's lives through the elimination of child labor, the conservation of civic beauty, the elimination of waste in education and war, the conservation of manhood, and the conservation of the Anglo-Saxon race![5] Despite the protests of those present who hoped to establish an organization to discuss practical problems, the Congress voted that its purpose would be "broad to act as a clearing house for all allied social forces of our time, to seek to overcome waste in natural, human, or moral forces."[6] The members refused to heed the advice of the president of the organization which had called the meeting,

[2] National Conservation Commission, *Report*, v. 3, pp. 620–751.

[3] *Proceedings, National Conservation Congress*, 1910, iv.

[4] *Ibid.*, 1913, 4. [5] *Ibid.*, 1909, 7–182. [6] *Ibid.*, 1909, 186.

when he warned, "If we try to cover the whole earth and heaven above and the future life with our conservation movement, we shall bite off more than we can chew. . . ."[7] Pinchot carried the meaning of conservation to its extreme limits, when in 1914 he campaigned as Progressive candidate for Senator from Pennsylvania on a platform calling for the conservation of human rights, natural resources, human welfare, and citizenship.[8]

This broad and sometimes vague emphasis both helped and hindered the cause of conservation. On the one hand it provided sorely needed popular support, especially from women's organizations and from the general public. But, it also created hostility among those who wished to deal with concrete problems in systematic fashion. One critic wrote Pinchot, in alarm, that the National Conservation Association had thrown open its doors to almost anything and everything which could be brought under the name of conservation. If it were more businesslike, and divested of irrelevant subjects, he concluded, he would be glad to be a member.[9]

Many in the National Conservation Congress considered this attempt to court a broad public sentiment as a political scheme to support the Insurgent revolt against Taft and Theodore Roosevelt's candidacy for the presidency. A Pittsburgh engineer complained that, while he had expected the 1910 Congress to accomplish something, instead he had found that he, as a delegate, was supposed merely to listen to speeches boosting former President Roosevelt and the New Nationalism.[10] A prominent New England conservationist wrote Pinchot early in 1912 that he felt the National Conservation Association to be "in great danger of becoming mixed up in the campaign to promote the political fortunes of some candidate for the presidency. . . ."[11] Despite Pinchot's desperate attempt to prove that the Association's directors included Democrats, regular Republicans, and Insurgents, many resigned from its ranks.[12] The secretary of the Los Angeles Chamber of Commerce wrote, in declining

[7] *Ibid.*, 1909, 185. [8] A copy of Pinchot's platform is in GP #1828.
[9] Thomas H. Johnson to Pinchot, Jan. 3, 1911 (GP #1819). [10] *Ibid.*
[11] Allen Chamberlain to Pinchot, Jan. 11, 1912 (GP #1818).
[12] Pinchot to John F. Bass, Dec. 23, 1912 (GP #1817).

membership, that, while the Chamber was nonpolitical and could not take sides in party matters, the National Conservation Association was closely linked with a partisan movement.[13] Pinchot's denials to the contrary, he thought of conservation in political terms. Roosevelt, he hoped, would be reelected president, and would place the former Chief Forester in charge of the administration's resource policies.

The American Forestry Association

Fully conscious of disagreements within the movement, in the summer of 1908 Pinchot set out to perfect an association to unify conservation sentiment to support measures in Congress. He formed the Conservation League of America, of which a Chicago lawyer, Walter L. Fisher, became president.[14] This loose coalition of twenty different established organizations, mainly waterway associations, proved to be ineffective. Each group concentrated on its own specific objectives and hesitated to spend time and money supporting other causes.

Pinchot next turned to the American Forestry Association which, with the National Irrigation Congress, had provided the major organizational backing for the Roosevelt resource policies. Since 1901 the Association had published *Forestry and Irrigation* to further the cause of federal measures in both fields. A few years later a congressional committee destroyed the effectiveness of the National Irrigation Congress by revealing that most of its operating funds came from Western railroads. At the same time, exponents of the Roosevelt administration's policies had since 1900 infiltrated the American Forestry Association. By 1908, therefore, Pinchot looked upon that Association as a logical vehicle for conservation agitation.[15] Under his influence, in the fall of that year the Association changed the name of its magazine to *Conservation* and broadened its scope

[13] H. B. Gurley to Pinchot, Jan. 26, 1911 (GP #1819).

[14] Pinchot to Charles R. Crane, June 4, 1908 (GP #2137); folder, "Conservation League of America," (GP #120).

[15] Pinchot to Curtis Guild, Jr., June 7, 1909 (GP #120).

to include many articles on matters other than irrigation and forestry.[16]

This change in emphasis satisfied neither the conservationists nor those interested more strictly in forestry. The former felt that forestry should play a secondary role in the broader movement, while the latter feared that conservation activity would swallow their program. Pinchot hoped to establish *Conservation* as the official organ of the National Conservation Commission, but feared that the management as of 1909 did not sympathize with the larger movement sufficiently to play that role.[17] At Pinchot's urging, the American Forestry Association appointed a new executive secretary who would stress a conservation viewpoint in both the magazine and the Association's affairs.[18] But the Chief Forester still did not trust the Association's officers to carry out his policies and within a few weeks he had enticed the new secretary to assume a similar position in his own organization—the National Conservation Association.[19]

Those who wished to stress forest management resented Pinchot's influence in the American Forestry Association. For example, in 1908 that body had concentrated its efforts on support for the Appalachian Forest measure then in Congress. Broader conservation issues took second place before this primary concern. Moreover, the organization had attracted a number of private and public forest officials who hoped to achieve cooperation among federal, state, and private groups for such specific improvements in forest management as fire prevention and control. These leaders successfully resisted Pinchot's attempt to transform the Association. After several joint meetings, the two groups decided to continue independently.[20] In January 1910 the name of the Association's journal was changed to *American Forestry*. At the same time its officials announced that in the future they would emphasize state and private forestry rather

[16] *American Forestry* (Jan. 1910), 16, 45.

[17] Henry S. Graves to Pinchot, Apr. 12, 1909; Pinchot to Graves, Apr. 17, 1909; Graves to Pinchot, Apr. 19, 1909 (all in GP #91).

[18] Pinchot to Graves, Apr. 17, 1909 (GP #91).

[19] *American Forestry* (Feb. 1910), 16, 93.

[20] *Ibid.*

than federal activities. The organization began increasingly to oppose Pinchot's point of view on resource matters. The Pinchot group, in turn, felt that the Association had "sold out" to private lumbermen who, having no interest in scientific forest management, merely used the national body as a public relations and publicity venture. Although the Association elected Robert P. Bass, a close friend of Pinchot's, as its president for 1911–13, two opponents of the former Chief Forester's expanding views, Professor Henry Drinker, and Charles L. Pack, headed the organization from 1913 to 1922. These changes in the Association entailed a shift from a broad program to more limited forest activities, and a departure from almost exclusive attention to a federal resource program to include state and private forest affairs as well.

The National Conservation Association

After his failure to capture the American Forestry Association, Pinchot set out in the summer of 1909 to form the National Conservation Association.[21] Disgusted with groups interested solely in their own special problems, Pinchot hoped to rely for support upon interested individuals throughout the country to whom he could appeal in terms of broad conservation principles rather than of specific self-interest.[22] Actually, the formation of the National Conservation Association narrowed rather than broadened the conservation movement. Although it did not work for limited forest or water development aims, the Association confined its activities to different, though equally specific, measures which Pinchot and others who defended the Roosevelt policies wished to stress. Since Pinchot's primary interests lay in water power and mineral leasing, the public identified the National Conservation Association with those issues, and for the next ten years congressmen referred to both questions as "the conservation problem." Moreover, when in the 1920's leaders of the Association wrote of their past accomplish-

[21] Pinchot to Charles R. Crane, June 4, 1908 (GP #2137).

[22] Pinchot to Walter Fisher, Feb. 23, 1909; Pinchot to Charles W. Eliot, June 7, 1909 (both in GP #120).

ments, they dwelt primarily on those two topics.[23] Due to Pinchot's predominance in resource matters, the public soon forgot that conservation had ever involved waterways, irrigation, grazing, or other issues vitally important in the Roosevelt resource program.

The National Conservation Association, moreover, confined its activities almost entirely to federal legislation. Pinchot had originally hoped that it would influence state as well as national policies,[24] but the organization soon lost touch with state activities. By 1912 it did not even have in its office a list of the names of the various state conservation bodies supposedly affiliated with it,[25] and eventually the public thought of the conservation movement as primarily an effort to influence national legislation.[26]

As the National Conservation Association's interests narrowed, its broad base of popular support, if it had any to begin with, rapidly declined. When Pinchot organized the Association he had estimated that its membership would rise quickly to 50,000 or 100,000.[27] But the legislative activity which the Association undertook did not attract a large following. Funds appeared from a few large donors rather than in small amounts from many members.[28] To bolster lagging support, in February 1911 Pinchot brought out the first issue of *American Conservation*, which seven months later discontinued publication. Although the official statement of the magazine's demise denied that it had failed to meet the expectations of its founders,[29] Pinchot wrote to one of the Association's directors that, "The difficulty of getting subscribers shows that a widespread

[23] Philip P. Wells memorandum, "National Conservation Association," in FN #10.

[24] Pinchot to Fisher, Feb. 23, 1909 (GP #120).

[25] Harry A. Slattery to Albert A. Hopkins, June 4, 1912 (GP #1817).

[26] Some state leaders in forest and other conservation matters complained bitterly of the concentration on national issues. See an account of a meeting of the American Civic Association of Nov. 19–22, 1907, written by Myra L. Dock in Dock MSS. See also Dock to J. Horace McFarland (*c.* Mar. 1908) in Dock MSS.

[27] Pinchot to Fisher, Feb. 23, 1909 (GP #120).

[28] Pinchot to John F. Bass, Dec. 28, 1911 (GP #1817).

[29] *American Conservation* (Aug. 1911), 1, 227.

popular demand for a magazine devoted to Conservation does not yet exist."[30] The editor had advertised the publication widely and had distributed 21,000 sample copies, but total subscribers had reached only 2,160 and the editor had failed to obtain paid advertising. Pinchot actually discontinued the venture with relief; he preferred to spend less time educating the public and more on the "real work of the Association"—legislative battles.

Much of the Association's increasingly limited character stemmed from Pinchot's strong influence in its councils. He became the organization's president in January 1910 and continued in that capacity until the group formally disbanded in the 1920's.[31] Many members did not relish his leadership; they welcomed the powerful support of his name and prestige, but opposed his domination. Moreover, because of the organization's identification with Pinchot and Pinchot's ties with Roosevelt, specific resource groups chose not to enlist its assistance in obtaining favorable legislation. Newlands and Maxwell, for example, turned down the Association's overtures of aid in securing action on the Newlands waterway bill because they feared that the measure would become involved in political dissensions and would make more enemies than friends in Congress.[32]

National Conservation Congress

The experience of the National Conservation Congress was similar to that of the National Conservation Association. The influence of the Pinchot group produced dissension within the ranks of the Congress, and finally disrupted it completely. Those who organized it in Seattle, Washington in September 1909, intended its annual sessions to provide an opportunity for state, federal, and private conservation leaders to share their ideas, their experiences, and their

[30] Pinchot to Clarence P. Dodge, July 29, 1911 (GP #1817).

[31] John F. Bass to Walter Fisher, Jan. 14, 1910 (GP #1817).

[32] Newlands' secretary to George Maxwell, Feb. 2, 1914; Maxwell to George Hudson (Newlands' secretary), Feb. 6, 1914 (FGN, Waterways-River Regulation, Correspondence, 1914, 1).

problems.[33] But Pinchot and others interested in federal problems, convinced that this approach involved a false brand of conservation, brought to Seattle a delegation sufficiently strong to capture the organization. During the next four meetings the Pinchot group effectively controlled its deliberations; the resolutions passed and the officers elected at succeeding sessions agreed substantially with the emphasis in conservation matters which the former Chief Forester stressed.

Pinchot hoped to absorb the National Conservation Congress into the National Conservation Association, but after protests from other Congress leaders, he abandoned these attempts.[34] Due to his efforts, however, future Congress meetings served not as a clearinghouse to discuss state and regional questions, but as an arena to debate national issues and as a voice to support national policies of the Pinchot-Roosevelt stamp. During the meetings from 1910 to 1912 those controlling the Congress appointed delegates, selected committees on resolutions, and named speakers who would favor a Roosevelt type of conservation.[35] As the water power question emerged predominant in the annual meetings and provoked considerable division of opinion, each succeeding Congress produced a more spirited debate and growing internal bitterness. Both sides picked delegations, tried to pack the resolutions committees, and used all the parliamentary tricks they knew.[36] Legislative maneuvering rather than an exchange of conservation ideas became the chief activity of the annual sessions of the Congress.

Struggle for control of the National Conservation Congress came

[33] *Proceedings, National Conservation Congress*, 1909. The Washington Conservation Association was organized by the Washington Forestry Association in November 1908; see *Ibid.*, xv.

[34] Bernard N. Baker to L. Frank Brown, Nov. 29, 1909 (GP #120); Pinchot to Baker, Nov. 12, 1909 (GP #1809).

[35] See the yearly proceedings on the Congress, 1909–1913.

[36] At the Fifth Congress in 1913 out of a total of 722 delegates, 169 were from the District of Columbia, mostly staunch Pinchot supporters. Although only 20 of them could vote, they could make noise from the floor to support Pinchot's views.

to a head in 1913 over the explosive water power issue.[37] Roosevelt administrators had formerly agreed to promote hydroelectric power installations as part of a comprehensive river development program so that proceeds from the sale of power could finance the entire scheme. But by 1913 several prominent engineers, previously associated with the Roosevelt program, had come to believe that the controversy hampered development and now supported a law more attractive to private promoters. Led by Marshall O. Leighton, Chief Hydrographer of the Geological Survey, George F. Swain, president of the American Society of Civil Engineers, and Lewis B. Stillwell, president of the American Institute of Electrical Engineers, they proposed legislation which would provide sufficient security and definiteness to attract private capital, and at the same time would play down the need for a federal charge for privately produced power.[38] For example, in 1908 Leighton had argued that charges for power could defray the costs of multiple-purpose development;[39] he now shifted to the view that the federal government should force profits to be passed on to the consumer in lower rates rather than appropriate them for its own use.[40]

Pinchot, Herbert Knox Smith, Commissioner of Corporations, and other former Roosevelt leaders, reacted sharply to this change. Forest Service policies, they argued, had not retarded development, and they could safely be applied to the navigable streams as well as to the public lands. Moreover, the power companies "should pay for what they get to reward the public" as a matter of principle, as well as help to finance broader multiple-purpose development.[41] The great danger, they said, stemmed from the ability of hydroelectric companies themselves to prevent progress by holding sites for speculation. In 1911, so the Bureau of Corporations had reported, the ten

[37] *Water Power Proceedings of the Fifth National Conservation Congress at Washington, D.C., November, 18, 19 and 20, 1913* (Washington, D.C., 1914), 11–23, 35–131.

[38] See argument of Frederick H. Newell, "Memorandum concerning water power presented to the Inland Waterways Commission, April 1908," (GP #2136); *Forestry and Irrigation* (April 1908), 14, 191.

[39] IWC, *Report*, 490.

[40] *Proceedings, National Conservation Congress, 1913*, 18.

[41] *Ibid.*, 41.

largest electric groups in the country controlled 3,270,00 horsepower, developed and undeveloped.[42] By 1913, this estimate had risen to 6,270,000 horsepower, and, according to Pinchot, the increase consisted mainly of sites held unused.[43] Pinchot and his friends, therefore, suspected private companies far more than did the three engineers, and stressed the dangers of monopoly rather than the need for working out a compromise agreement.

These differences burst forth at the fifth National Conservation Congress in 1913. President Charles L. Pack had appointed a special water power committee, including the three engineers and Pinchot, to report at the meeting. Pinchot looked upon this procedure as a device to secure endorsement by the Congress of his own water power views, and he was alarmed to discover that Swain, the chairman of the committee, was prepared to make concessions, even to the extent of eliminating charges.[44] The original Swain draft stressed the lack of development, and the final report, which he prepared, announced, "we believe that in many cases the real interests of conservation will be best served by making no government charge. . . ."[45] In working on the preliminary draft, Pinchot realized that a majority of the committee would oppose him. He, therefore, prepared to present a minority report to the Conservation Congress which would emphasize monopoly and spell out the need for charges and the precise restrictions on the operations of private companies.

The Congress' proceedings, held at Washington, D.C., witnessed a vigorous jockeying for position on both sides. The galleries favored Pinchot, for he had packed the hall with sympathizers. In the face of protests from the chairman of the convention, Charles

[42] Bureau of Corporations, *Report on Water Power* (Washington, 1912). See also 62nd Congress, 2nd Session, National Waterways Commission, *Hearings on the Development and Control of Water Power, Sen. Doc. #274.*

[43] *Proceedings, National Conservation Congress, 1913*, 35.

[44] Pinchot to Garfield, Oct. 23, 1913; Pinchot to Swain, Oct. 31, 1913 (JRG, Cabinet #3, Tray #3, Misc. Corr. 1911–18). Pinchot to Stimson, Oct. 31, 1913; Pinchot to Herbert K. Smith, Oct. 31, 1913; Smith to Pinchot, Nov. 13, 1913; Pinchot to Smith, Nov. 14, 1913 (all in GP #1826).

[45] For the Swain report see *Proceedings, National Conservation Congress, 1913*, 10–23.

L. Pack, Pinchot presented his minority report and secured its adoption over the majority Swain report. Still smarting at what he felt was a betrayal by the engineers, Pinchot refused to stop here, and the following day, after many had gone home on the assumption that the fighting was over, he rubbed it in on the opposition by obtaining approval of a special resolution again emphasizing the danger of a water power monopoly.[46]

As a result of these events, two leaders of the Roosevelt conservation movement, Pinchot and Leighton, parted company. On May 1, 1913, Leighton had resigned his post as Chief Hydrographer in the Geological Survey to enter private engineering consulting as a specialist in water power. Prior to that time, on April 11, Pinchot had expressed approval of Leighton as a member of the water power committee for the Conservation Congress.[47] But he soon changed his mind. In an exchange of correspondence early in November, the former Chief Hydrographer pleaded with Pinchot, "Surely, we who have been so thoroughly in sympathy in a common and great cause . . . are in duty bound to try to effect an agreement in the present juncture."[48] When Leighton invited the former Chief Forester to "talk this over,"[49] Pinchot replied simply that he would submit a minority report.[50] Differing with his former associate at the Congress, Leighton announced, "I know I am taking the unpopular side, and I know that I am breaking with a man whom I have followed and loved and assisted to the best of my ability for more than ten long years."[51] Pinchot replied that he did not consider Leighton to be "on our side any longer."[52] From this time on Pinchot and his friends spoke of Leighton in scornful tones.[53]

The 1913 controversy ended all possibility of a successful compromise on the water power issue. Such compromise, in early 1913, seemed promising, for some corporations were encouraging their

[46] *Ibid.*, 67–131. [47] Pinchot to C. L. Pack, April 11, 1911 (GP #1828).
[48] Leighton to Pinchot, Nov. 1, 1913 (GP #1824).
[49] Leighton to Pinchot, Nov. 12, 1913 (GP #1824).
[50] Pinchot to Leighton, Nov. 14, 1913 (GP #1824).
[51] *Proceedings, National Conservation Congress, 1913*, 119.
[52] P. S. Stahlnecker to Slattery, Feb. 28, 1913 (GP #1826).
[53] Slattery to Stahlnecker, July 29, 1915 (GP #1833).

colleagues to accept charges.[54] During the preceding two years, Secretaries Stimson and Fisher had conducted many satisfactory talks with hydroelectric officials. Stimson had perfected a model bill for the Connecticut River, a bill which the power companies approved and which included charges, while Fisher had developed his permit system for the public domain without much resistance. Some hydroelectric corporations favored national regulations because they preferred one system rather than a multitude of different state laws.[55] But the Conservation Congress sharpened the controversy beyond all possibility of immediate agreement. There was now no middle ground. To Pinchot, Swain had now "turned completely over."[56] He wrote a friend in Boston to determine if Swain had financial interests in water power development, adding, "Unless he [does], it is a little difficult to see how he can have so intimate a knowledge of exactly what the waterpower interests desire."[57] By January 1914, the private corporations also had changed their attitude considerably, and displayed little desire to carry through negotiations.[58]

The 1913 Congress session also brought to a head dissatisfaction with the Chief Forester's personal influence in the organization. Those who had prevented his attempt to absorb the Congress into the National Conservation Association now looked on with disgust as the Congress became a debating society on national issues and neglected more concrete local problems.[59] They especially criticized Pinchot for insisting on thrashing out the water power issue each year. Since many of them were not vitally interested in water power, they preferred to soft-pedal that issue so that it would not disrupt the proceedings and prevent attention to other problems. Torn

[54] Wells to Slattery, Sept. 22, 1915 (GP #1834).

[55] Statement by John D. Ryan of the Montana Power Co. in Wells Memorandum, Aug. 5, 1913 (GP #1666).

[56] Pinchot to Garfield, Oct. 23, 1913 (JRG, Cabinet #3, Tray #3, Misc. Corr. 1911–18).

[57] Pinchot to Ernest E. Smith (c. Oct. 1913) (GP #1826).

[58] Wells to H. K. Smith, Jan. 22, 1914 (GP #1668).

[59] E. T. Allen to Dr. Henry S. Drinker, Aug. 4, 1916 (GP #1829); E. T. Allen to Dr. G. E. Condra, July 31, 1916 (GP #1829).

by these differences, the Congress did not meet in 1914. Two years later the anti-Pinchot faction called a Congress meeting in Washington at which they hoped to oust their opponents; this Congress adopted resolutions favoring water power measures which Pinchot opposed, dropped the term "conservation" from the Congress' statement of purposes, and elected officials not particularly friendly to the Chief Forester.[60] A letter to Pinchot from the new president of the Congress revealed the bitterness involved. Since Pinchot had in essence informed them that he would not support the Congress unless it was padded according to his views, wrote the president, the new executive committee preferred to receive no money from the Forester to publish the Congress' proceedings and to dissociate itself entirely from his leadership.[61]

A year later Pinchot again obtained control of the organization. He became its president by agreeing that the water power issue would never again be raised in the annual meetings.[62] The dissidents also demanded that he refuse to bind the organization one way or the other on water power by declining to support or oppose measures on this question;[63] but this stipulation the Chief Forester would not accept. He interpreted his selection as committing the group to his principles, and wrote to a member of the executive board that he stood squarely against the Shields and Myers water power bills, endorsed by the Congress against his will the year before, and would do all he could to defeat them.[64] Despite tentative plans to hold a session of the Congress in 1917, Pinchot could not bring the group together again.[65] His adamant position on water power and his view that the Congress should function primarily to support that position wrecked the Conservation Congress and left a legacy of bitterness against him which took years to eradicate.

[60] Thomas Shipp to members of the National Conservation Congress, nd (GP #1833); Slattery to Treadwell Cleveland, May 9, 1916 (GP #1829).

[61] Condra to Pinchot, Sept. 11, 1916 (GP #1830).

[62] J. B. White to Harry Slattery, Sept. 25, 1916 (GP #1825).

[63] Henry S. Drinker to George E. Condra, Oct. 3, 1916 (GP #1830).

[64] Pinchot to Drinker, Oct. 11, 1916 (GP #1830).

[65] Pinchot to William K. Kavanaugh, Dec. 20, 1916 (GP #1831).

Preservation versus Conservation

Struggles within the organized conservation movement gradually narrowed its emphasis to the Pinchot water power policy. An equally bitter conflict persisted between "conservationists" and "preservationists," between those who favored resource development and others who argued that wild areas and wildlife should be preserved from commercial use. The latter, including arboriculturists, natural historians, and such sportsmen's organizations as the Boone and Crockett Club, derided "forestry for profit"; they were concerned less with a rational timber policy and more with preservation of trees and wildlife as objects of beauty, scientific curiosity, and recreation.[66] This difference in outlook pervaded a great number of resource incidents during and after the Roosevelt administration, and led to mutual suspicion, scorn, and distrust. Each group claimed the banner of true conservation and accused the other of being false standard bearers of the gospel.

Aided by President Roosevelt, himself a natural historian of no mean ability and an enthusiastic sportsman, preservationists accomplished much during the early years of the century. C. Hart Merriam, the first director of the Biological Survey, and Roosevelt persuaded Congress to raise that agency to Bureau status and to place federal wildlife work on a firm basis.[67] The Society for the Preservation of Historical and Scenic Spots, organized in 1900, worked for a law which would authorize the chief executive to set aside such areas as permanent reserves. Finding an indefatigable advocate of its cause in John D. Lacey of Iowa, Chairman of the House Public Lands Committee, the Society finally secured its objective in the Antiquities Act in 1906. With the advice of the Audubon Society and related organizations, Roosevelt established throughout the country a host of bird sanctuaries.[68] In 1913 Congress

[66] Martha A. Moffet to Myra L. Dock, Sept. 29, 1900 (MLD).

[67] Roosevelt to James W. Wadsworth, Feb. 18, 1904 (TR).

[68] For historical background on wildlife activities see William T. Hornaday, *Our Vanishing Wild Life* (New York, 1913); Hornaday, *Thirty Years War for Wild Life* (New York, 1931); Theodore B. Palmer, *Chronology and Index of the More Important Events in American Game Protection, 1776–1911, United States Department of Agriculture, Biological Survey, Bulletin #41.*

provided for migratory bird refuges and for penalties for killing migratory birds.[69] Moreover, state fish and game laws marked the steady growth of a more rational system of fish and game protection,[70] and Congress granted authority to federal forest reserve employees to aid in administering these laws in the national forests.

Yet, preservationists obtained this impressive list of measures without the assistance of the Pinchot-Garfield group. They maintained few close contacts with the Roosevelt administration's resource planners and their allied organizations, and in fact frequently opposed them.[71] Prior to Pinchot's emergence as a conservation leader, preservationists had exercised considerable influence in resource affairs. Visualizing the Forest Reserve Act of 1891 as a step toward reserving resources from development, they had worked to increase the number of forests, to restrict their use, and to police them more effectively. They had considered the Act of 1897 as an opportunity for protecting the reserves from commercial exploitation and not as an entering wedge for development.[72] Pinchot disagreed violently with this viewpoint. He tried to counteract it by empha-

[69] Harry A. Slattery to Mr. Haskell, Sept. 8, 1915 (GP #1830); proclamation of the President, "Regulations for the Protection of Migratory Birds" (GP #1706); Report of the Committee on Conservation of Forests and Wildlife of the Camp Fire Club of America (GP #1671).

[70] For New York State, as an example, see Gurth Whipple, *Fifty Years of Conservation in New York State, 1885–1935* (New York, 1935), 101–156.

[71] J. Horace McFarland to Pinchot, Nov. 26, 1909 (GP #1809). In the late 1920's Frederick H. Newell was engaged in revising Van Hise's book on conservation. In a memorandum commenting on the State Conservation Commissions, in this connection, he wrote, "Most of these had not given full consideration to the larger questions of natural resources but confined their attention mainly to State parks, wild life and recreational features" (FN #10). With respect to the post-1908 conservation period, Newell wrote to Pinchot, "My general impression is that after the first enthusiasm died down the general attitude was not particularly favorable and there was a tendency to divert the conservation movement into narrow channels of fish and game protection." Newell to Pinchot, May 28, 1928 (FN #10).

[72] For this attitude toward the reserves and the controversy it aroused in the Pacific Northwest, see Lawrence Rakestraw, "A History of Forest Conservation in the Pacific Northwest, 1891–1913," Ph.D Thesis, University of Washington, 1955, pp. 28–124.

sizing a distinction between scientific forestry and preservation,[73] and by opening the forests to grazing. The transfer of the reserves to the Department of Agriculture in 1905 represented the victory for the development point of view in the Roosevelt administration. The change of name from "forest reserves" to "national forests" symbolized its significance.

The running battle over the New York Adirondack State Park further revealed the conflict between these two groups.[74] Preservationists had induced New York voters to include a provision in the state constitution of 1894 prohibiting all timber cutting in the Park. The two major groups behind this move—the New York Board of Trade and Transportation, which feared a loss of water supply for the Erie Canal, and the Association for the Protection of the Adirondacks, dominated by owners of private estates within the Park area—persuaded the state to strengthen this provision from time to time and to reaffirm it in the constitution of 1915. Originally they had argued that the restriction would be only temporary and would be lifted when danger of wanton lumbering ceased;[75] but they later adopted the viewpoint that the lands should remain "forever wild," and should never be cut even by the state.[76]

Pinchot and other foresters argued that the constitutional limitation was too broad and ill-advised.[77] The state could not cut mature timber, construct roads to make areas accessible for fire protection, assure reproduction stands, or carry out a general program of sustained-yield timber management. Spurred on by this sentiment, the New York Assembly agreed to a proposal to draw up a timber management plan for the Adirondacks, and in 1900 the State Forest Commission requested the U. S. Division of Forestry to present a

[73] Pinchot, *Breaking New Ground*, 71.

[74] For history of the Adirondacks see Whipple, *Fifty Years*, and Alfred L. Donaldson, *A History of the Adirondacks* (2v, New York, 1921). Both are very inadequate, failing to deal with the conflicting forces involved in the bitter Adirondacks struggle of over fifty years.

[75] Donaldson, *Adirondacks*, II, 193, 255.

[76] For an expression of this opinion see editorial, "Forever Wild," in *New York Times*, Nov. 23, 1951.

[77] MSS, "Adirondack plan," dated Feb. 12, 1895 in GP #1811.

proposal for a selected township.[78] On protests from Adirondack residents the undertaking was dropped.[79] Pinchot then turned to other devices. To demonstrate to the public that private owners could manage their lands well he cooperated with the latter to establish experimental management tracts.[80] To counteract the influence of the private estates and clubs within the Park he recommended that all privately owned land inside the boundaries—60 per cent of the total—be condemned and the state's holdings be consolidated. In 1911 he worked through the Camp Fire Club in an effort to obtain a state law advocating scientific forestry in the Adirondacks.[81] But these efforts failed. The Board of Trade and Transportation and the Association for the Preservation of the Adirondacks thwarted each move, and prevented scientific forestry in the Park.[82]

While the Adirondacks aroused little public interest outside New York State, the controversy over the Hetch-Hetchy Valley of California stirred the entire nation. John Muir's Sierra Club had successfully persuaded Congress to incorporate this recreation spot of rare beauty into Yosemite National Park.[83] The Valley also contained a valuable reservoir site, which San Francisco desired to use for its water supply. Secretary of the Interior Ethan A. Hitchcock, who favored the preservationist view, reacted coolly when the city applied to the federal government for permission to use the area. But his successor, James R. Garfield, approved the project. Thoroughly aroused, Hetch-Hetchy defenders argued that San Francisco could

[78] Pinchot, *Breaking New Ground*, 182–185.

[79] The reaction against scientific forestry in the Adirondacks came over the issue of the Cornell College of Forestry. See Donaldson, *Adirondacks*, 2, 201–206; *Forest and Stream* (Dec. 28, 1901), 57, 502.

[80] Overton W. Price to Pinchot, Nov. 1, 1911 (GP #1817).

[81] William Edward Coffin to Pinchot, Dec. 8, 1911; Pinchot to Coffin, Dec. 12, 1911; Coffin to Pinchot, Dec. 14, 1911; "Minutes of meeting of Second Conference on the Adirondacks," Dec. 19, 1911 (all in GP #1818).

[82] The two groups also came into conflict over the development of water power in the Adirondacks. See undated and unsigned letter to Pinchot in GP #1817.

[83] Robert Shankland, *Steve Mather of the National Parks*, 47–49.

tap other sources; the city replied that it could not. Each side contended that it represented the "public interest," while its opponents spoke for "private interests." Yet, conflict between two public uses of the Valley was the crux of the controversy. Believing that water supply comprised a more important public use of the area than did recreation, Pinchot gave Garfield his full support and persuaded President Roosevelt to do likewise.[84] The issue became so bitter that John Muir was not invited to the Governors' Conference in 1908.[85] Wrangled over in Congress for almost a decade, in 1914 a bill authorizing the reservoir finally passed.

Preservationists fought back bitterly against the administration. John Muir tried desperately to influence his friend, President Roosevelt, who, torn between the two uses, supported Pinchot only with indecision.[86] The women's clubs, staunch defenders of the federal forest policy and of Pinchot's entire program, could not understand his view in this case. Following their more basic distrust of "commercialism" of all kinds, they vigorously opposed the reservoir.[87] Many "preservationists" wrote the National Conservation Association to resign from a group led by false prophets.[88] The entire "Hetch-Hetchy steal" remained for years as a warning to park enthusiasts. The affront to Muir in 1908, his "most dispiriting setback," was bitterly resented by the aging woodsman and park and wilderness groups long remembered the event with equal ill-will. Muir vented his indignation to the secretary of the Sierra Club, "Never mind, dear Colby, the present flourishing triumphant growth of the wealthy wicked, the Phelands (sic) [mayor of San Francisco], Pinchots and their heirlings, will not thrive forever. . . .

[84] Pinchot to Allen Chamberlain, Aug. 15, 1909 (GP #121); Slattery to C. L. Pack, July 3, 1913 (GP #1827).

[85] Shankland, *Steve Mather*, 51.

[86] Roosevelt to Ethan A. Hitchcock, Feb. 8, 1905; Roosevelt to John Muir, Sept. 16, 1907; Roosevelt to Garfield, Apr. 27, 1908; Roosevelt to Muir, Apr. 27, 1908; Roosevelt to Robert U. Johnson, Dec. 17, 1908 (all in TR).

[87] Harry Slattery to William Kent, Oct. 17, 1913 (GP #1823).

[88] W. B. Bourn to Pinchot, Jan. 11, 1911 (GP #1818); Allen Chamberlain to Pinchot, Jan. 11, 1912 (GP #1818); G. Frederick Schwartz to the National Conservation Association, June 9, 1913 (GP #1826); H. Meyer to Slattery, July 28, 1913 (GP #1825).

We may lose this particular fight, but truth and right must prevail at last."[89]

This opposition appalled Pinchot, Newell, and Garfield. Harry Slattery, secretary of the National Conservation Association, wrote to Garfield in 1913, "Unfortunately, our good friends the nature lovers are still unreasonable in their attitude. There is grave danger they will again be able to block this most necessary legislation."[90] Congressman William Kent of California wired Pinchot of a conspiracy "engineered by misinformed nature lovers and power interests who are working through the women's clubs."[91] Slattery confided to Charles L. Pack, of the American Forestry Association, "Some of our scenic friends have certainly let their heart run away with their head in making veiled insinuations that men like . . . Mr. Pinchot, Mr. Garfield, Secretary [of the Interior] Lane and others are being used by private interests on this Hetch-Hetchy question."[92]

The Hetch-Hetchy controversy severed relations between the leaders of the two rival movements, Gifford Pinchot and J. Horace McFarland, president of the American Civic Association. Since its formation in 1905, this organization had spearheaded the drive for municipal, state, and national parks.[93] Yet Pinchot had dealt gingerly with it. Although he wished to secure the support of its large following, he disagreed with its aesthetic emphasis. In 1909 he refused to merge it with the National Conservation Association.[94] Pinchot wrote McFarland during the Hetch-Hetchy controversy that year that the aesthetic side of conservation could not "at this stage of the game . . . go ahead of the economic and moral aspects of the case."[95] McFarland retorted, "I feel that the conservation movement is now

[89] Shankland, *Steve Mather*, 48–49.

[90] Slattery to Garfield, July 28, 1913 (GP #1827).

[91] Kent to Pinchot, Oct. 8, 1913 (GP #1823).

[92] Slattery to C. L. Pack, July 28, 1913 (GP #1828).

[93] Shankland, *Steve Mather*, 52–54.

[94] Pinchot to Charles W. Eliot, Aug. 5, 1909 (GP #120); John F. Bass to Walter Fisher, Nov. 9, 1909 (GP #1817); J. Horace McFarland to Pinchot, Nov. 22, 1909 (GP #1809); Pinchot to McFarland, Nov. 24, 1909 (GP #1809); Garfield to Pinchot, Jan. 5, 1910 (GP #121).

[95] McFarland to Pinchot, Nov. 26, 1909 (GP #1809).

weak, because it has failed to join hands with the preservation of scenery. . . . Somehow we must get you to see that the man whose efforts we want to conserve produces the best effort and more effort in agreeable surroundings; that the preservation of forests, water powers, minerals and the other items of national prosperity in a sane way must be associated with the pleasure to the eye and the mind and the regeneration of the spirit of man."[96] The two leaders, each feeling that the other had not looked fully or fairly into the Hetch-Hetchy case, ceased to correspond. McFarland wrote to a friend, "I will admit that I am not big enough, or wise enough, or strong enough to work wholeheartedly with anyone who has assumed a wrong position in respect to an important public matter in which I am interested, and declined to argue it out with me fairly, fully and frankly."[97]

The Hetch-Hetchy issue accurately reflected the views of both Pinchot and Garfield that national parks in general should be opened for such resource development as grazing and lumbering. Garfield proposed that dead and decayed timber be cut in the national parks.[98] The Forest Service, in turn, opposed bills to create new parks which did not provide for development. When a measure to create Glacier National Park appeared in Congress, prohibiting all commercial use save removal of dead, down, or decaying timber by settlers, the Forest Service prepared a rival measure which would permit cutting of mature timber, water power development, and railroad construction within the Park.[99] The measure to set aside the Calevaras Big Trees as a national park prompted a similar counter-proposal from Pinchot, to permit cutting of any timber in the park

[96] *Ibid.* [97] McFarland to Overton Price, Mar. 31, 1913 (GP #1825).

[98] Herbert Parsons to Pinchot, Jan. 24, 1910; Pinchot to Parsons, Feb. 10, 1910 (GP #1819).

[99] Memorandum for Pinchot, Nov. 8, 1907; memorandum for Pinchot, Dec. 23, 1907; memorandum, Wells to Pinchot, Mar. 28, 1908; Price to George Bird Grinnell, Apr. 7, 1908; Pinchot to Senator Carter, Apr. 7, 1908; memorandum for Woodruff, Apr. 9, 1908 (all in RG #95, Law Office, Correspondence files).

except the Big Trees themselves.[100] When competition arose between rival groups for inclusion of a particular area in a national forest or a national park, Pinchot usually supported the former. For example, in 1916 he helped to extend the Rocky Mountain National Forest in Colorado to an area which preservationist groups hoped would become part of Rocky Mountain National Park.[101]

As early as 1904 Pinchot recommended that Congress transfer the national parks to the Forest Service, so that he could administer them according to these views. Garfield, as Secretary of the Interior, supported the proposal. Bills to accomplish this purpose, prepared by Pinchot, met vigorous opposition from preservationists. Their spokesman in the House, Representative John F. Lacey, Chairman of the House Public Lands Committee, squelched the measure in 1906 and 1907. After that date, park advocates gathered increasing strength outside Congress which enabled them to thwart the Pinchot proposal and at the same time to take the offensive in 1910 by demanding that Congress create a National Park Bureau in the Department of the Interior. Travel agencies, railroads, and highway associations, all interested in a swelling Western tourist trade, joined with recreationists and sportsmen to back this counter measure.[102] Pinchot vigorously opposed it, arguing that a National Park Bureau was "no more needed than two tails to a cat."[103] By 1916 he agreed to support the proposal if the Bureau were placed in the Department of Agriculture,[104] and the Forest Service officially supported his view.[105] The park groups, however, won this round

[100] Pinchot to Senator George C. Perkins, Feb. 18, 1908 (RG #95, Law Office, Correspondence files).

[101] Slattery to Pinchot, Apr. 4, 1916 (GP #1834); Charles L. Hover to Pinchot, Sept. 11, 1916 (GP #1831).

[102] Shankland traces the development of the National Park Service in *Steve Mather*, 42–113.

[103] Pinchot to Garfield, Nov. 22, 1911 (JRG, Cabinet #3, Tray #3, Misc. Corr., 1911–18). See also J. Horace McFarland to Frederick Law Olmsted, Dec. 18, 1912 (GP #1819); Olmsted to Pinchot, Dec. 23, 1912 (GP #1819); Pinchot to Olmsted, Dec. 16, 1912 (GP #1819).

[104] Pinchot to Slattery, Apr. 6, 1916 (GP #1834).

[105] Secretary of Agriculture David F. Houston to Representative Scott Ferris, Feb. 12, 1916 (GP #1834).

when in 1916 Congress created a separate National Park Service in the Interior Department.[106]

This affair only confirmed the suspicions of Pinchot and McFarland. The latter feared that "the Forest Service as it is now controlled is inimical to the idea of a separate and adequate management of the National Parks,"[107] and to his friend, Frederick Law Olmsted, deplored Pinchot's stand on both the Hetch-Hetchy and national park issues: "I confess . . . that I have never had a harder jolt in my life than that which has come to me through his attitude on these questions, in which he has not been willing to admit error when shown."[108] For a short time Olmsted served as an intermediary between the two men. Pinchot wrote him that he opposed a separate Bureau of National Parks, that Western parks could not be handled like city parks as McFarland seemed to think, but only as vast areas with problems similar to the national forests. The former Chief Forester added that he had stopped discussing such matters with McFarland because it was a waste of time. "I could not get my views across to him," he wrote.[109]

Gifford Pinchot's role in these controversies deserves special attention. In later years preservationists erroneously utilized his name as a symbol of their aspirations. One can find few widely publicized pronouncements by Pinchot of his position. Yet the record of his private statements and actions clearly reveals hostility to those whom his friends derided as "nature lovers." He did not want to alienate them entirely, for he needed their support on other matters, but behind the scenes he worked quietly to offset their influence. For example, when the Chairman of the Outdoor Art League of California, Mrs. Lovell White, asked him to intervene with Roosevelt and Speaker of the House Cannon in favor of the proposed Calevaras Big Tree reservations, the Chief Forester replied through an aide, "Mr. Pinchot thinks that in view of his official position he cannot with propriety actively forward the interests of

[106] Peffer, *Public Domain*, 175–179.
[107] McFarland to Frederick L. Olmsted, Dec. 18, 1912 (GP #1819).
[108] *Ibid.*
[109] Pinchot to Olmsted, Dec. 26, 1912 (GP #1819).

proposed or pending legislative measures."[110] His vigorous activity on Capitol Hill in many other matters branded such a statement as a polite note of opposition. Secretary of the Interior Ballinger, in fact, was more friendly to the preservationists than was Pinchot. Ballinger took their side on the Hetch-Hetchy issue; in his annual report of 1910 he strongly supported a Bureau of National Parks;[111] and as Commissioner he fought to exempt the national parks from easements under the Right-of-Way Act of 1901 "in order that they may be preserved," he wrote, "for the purposes for which they were created."[112]

One must also reassess Pinchot's wider role in organized conservation affairs. Without question the Chief Forester contributed more than any other individual to public awareness of forestry and water power problems. He firmly planted the idea of conservation in the minds of the American people; he built up the United States Forest Service as a highly effective government agency and almost personally staved off measures which would have granted public utility corporations unlimited franchises. Yet, Pinchot also helped to retard the movement. His vigorous attempt to direct conservation into those limited channels he preferred to stress, and his refusal to compromise with those with whom he differed played a large role in splintering conservation organizations, contributed to conflicts among resource groups and to personal bitterness among their leaders, and alienated many who hesitated to become involved in the tense atmosphere surrounding such a controversial figure. A vigorous attack on resource problems other than national forestry, water power, and mineral leasing affairs suffered from his dominant and inflexible influence.

[110] Philip Wells to Mrs. Lovell White, May 4, 1908 (RG #95, Law Office, Correspondence Files).

[111] *Annual Report of the Secretary of the Interior, 1910*, 58, 61–62.

[112] Ballinger report on HR 3907 in RG #49, Bill Book #29, pp. 174–175.

Chapter X

The Corps of Engineers Fights Water Conservation

Conflicts within the organized conservation movement prevented Pinchot and Garfield from forming united backing for the Roosevelt policies. These leaders also met severe opposition from other interest groups who feared that they would have little opportunity to influence a more integrated resource administration. The fate of the proposed comprehensive water development program best illustrates these obstacles. For some ten years the measure to establish a multiple-purpose water policy, which Senator Francis G. Newlands first outlined in 1907, remained before Congress for approval. Senator Newlands and his able propagandist, George H. Maxwell, spearheaded the fight for the measure, but Pinchot and the National Conservation Association, although more concerned with other matters, approved the plan and at times entered the fray.[1] In 1916 Pinchot hoped to hold a special waterway congress which, under his own auspices, would again bring the recommendations of the Inland Waterways Commission to public attention.[2] Yet these efforts failed, and multiple-purpose river development did not achieve its first success until Congress authorized comprehensive river basin investigations for the entire country in 1927 and the following year approved the Boulder Dam Act.

[1] Newlands' secretary to George Maxwell, Feb. 2, 1914 (FGN, Waterways-River Regulation, Correspondence, 1914, 1); Pinchot to John F. Bass, Feb. 14, 1914 (GP #1827).

[2] Pinchot to William K. Kavanaugh, Dec. 20, 1916 (GP #1831).

The Corps Helps Defeat the Newlands Bill

The Corps of Engineers was especially alarmed over the Newlands measure. Its members feared that an integrated water program would transfer many of its functions to other agencies. The Corps bitterly fought the Newlands program throughout its history and at times the opposition of its officers was crucial. Senator Newlands hoped that his measure would create "teamwork on the part of the Engineer Corps of the Army and the other scientific and constructive services of the National Government that [related] in any way to water. . . ."[3] From the point of view of the Corps, however, as one prominent officer declared in October 1907, there was "great difficulty in bringing [federal water agencies] 'into coordination' without bringing them into subordination. . . ."[4] A junior officer of the Corps warned his Chief that the proposed waterway commission would "begin by hampering the Corps with dictations partaking more and more of the nature of orders, and [might] end by usurping some or all of its civil functions. . . . This would be a great misfortune to the Corps and to its members personally. . . . There would be friction between the Engineer Department and the Commission, and I believe that every Engineer Officer should oppose the enlargement of its power."[5]

Well aware of this attitude, Maxwell and Newlands tried to present their proposals so as not to antagonize the Corps. Early in 1911 Pinchot cautioned Maxwell, who was preparing a new draft of the bill, that he should take "some care to secure the Army Engineers their present job untouched."[6] At the same time, Brigadier-General Bixby, Chief of the Corps, expressed concern to Newlands over who would be chairman of the proposed commission.[7] Newlands agreed that the Chief of the Corps would hold this

[3] Newlands to A. L. Crocker, May 20, 1912 (FGN, Waterways-River Regulation, Correspondence, 1912, 2).

[4] Brigadier-General O. W. Ernst to Brigadier-General Alexander Mackenzie, Nov. 21, 1907 (RG #77, #62743).

[5] Capt. P. S. Bond to Mackenzie, Jan. 17, 1908 (RG #77, #62743).

[6] Pinchot to Maxwell, Feb. 15, 1911 (GHM, Clippings, Newlands Bill).

[7] Newlands to Maxwell, Mar. 2, 1913 (FGN, Waterways-River Regulation, 1912, 1).

position. When Bixby pointed out that the bill did not make this clear, the Senator assured him, "If there is any doubt upon this score, I will have the bill amended in this particular."[8] Newlands and Maxwell sought to avoid interference with existing programs administered by the Corps. Maxwell wrote the Senator that, in particular, the Mississippi River Commission should be supported fully in order not to antagonize the Corps.[9] Finding that many congressmen strongly opposed a commission in which the Corps would be outnumbered, Newlands, in turn, warned Maxwell, "I am sure that it [the bill] will not have the friendly support of the Engineer Corps of the Army until a change in this particular is made, and the influence of that Corps is very great."[10]

The success of the Corps in opposing the Newlands measure varied with the attitudes of the respective Secretaries of War. Secretaries Henry L. Stimson, under Taft, and Newton D. Baker, under Wilson, both heartily endorsed the Newlands plan. During their terms of office, the Corps could present its views to Congress only indirectly and not through official communications. More sympathetic secretaries, such as Lindley M. Garrison, during the first Wilson administration, enthusiastically conveyed the Corps's point of view to congressmen and to fellow cabinet members. Moreover, in 1917 and 1918 Secretary Baker, preoccupied with his war duties, gave little attention to water problems, and permitted the Corps itself to establish the official position of the War Department on the Newlands measure.

In 1908 Brigadier-General Mackenzie had disagreed with the report of the Inland Waterways Commission and with the Commission's approval of the Newlands bill. At that time President Roosevelt had taken special steps to counteract the opposition of both the Corps and Secretary of War Taft to comprehensive

[8] Newlands to Brigadier-General Bixby, Mar. 1, 1912 (RG #77, #83879/4).

[9] Maxwell to Newlands, Apr. 5, 1912 (FGN, Waterways-River Regulation, Correspondence, 1912, 2).

[10] Newlands to Maxwell, July 19, 1912 (FGN, Waterways-River Regulation, Correspondence, 1912, 3).

waterway legislation. When the Newlands bill came up again during 1912, however, Secretary of War Stimson, who favored it, used his influence to try to secure its passage. The Corps, in turn, endeavored to circumvent the Secretary, and to thwart the measure. As early as January 5, 1912 the Senate Committee on Commerce requested from the Secretary of War a report on the Newlands proposal. Stimson, in accordance with War Department routine, referred the request to the Chief of the Corps.[11] The Corps returned no reply. On March 14 the Committee again requested a report, but again the Corps remained silent.[12] A third request on December 17, 1912 brought action.[13] On December 30 the Chief of the Corps forwarded the request to the Board of Engineers for Rivers and Harbors with the notation, "For its views and *early* report."[14] On March 10, 1913, six days after Congress had adjourned, the Board submitted its "early" report, with its adverse recommendation:

> The Board . . . is unanimously of the opinion that its [the Newlands measure] adoption is undesirable in the interests of river improvement. . . . The bill is open to so many objections that the Board is compelled to regard it as impracticable as a means of improving streams for the uses of commerce and navigation, and is unable to suggest any modifications which would make it acceptable. . . . The consideration of terminal and transfer facilities, waterpower, and such other subjects as may be properly connected with projects for improvement of navigation is provided for by existing law . . . and these matters receive due attention in all cases where they can be advantageously coordinated with improvements for the benefit of commerce and navigation.[15]

[11] Secretary, Committee on Commerce to Secretary of War, Jan. 5, 1912 (RG #77, #83879).

[12] Secretary, Committee on Commerce to Secretary of War, Mar. 14, 1912 (RG #77, #83879/6).

[13] Secretary, Committee on Commerce to Secretary of War, Dec. 17, 1912 (RG #77, #83879/11).

[14] Chief of Engineers to Board of Engineers, Dec. 30, 1912 (RG #77, #83879/11); italics by the Chief of the Corps.

[15] Col. William T. Russell, Senior Member, Board of Engineers, to Chief of Engineers, Mar. 10, 1913 (RG #77, #83879/18).

In the meantime, on January 27, 1913, Secretary Stimson had sent to the Senate Committee a report favourable to the bill.[16] He had referred a first draft of this letter to the Corps for its comment. The Engineer Department returned the draft with marginal notations which denied the existence of new knowledge or new conditions in water development of which the Corps did not take cognizance, and disagreed with the view that federal water resource agencies had failed to cooperate.[17]

At this session of Congress, the Senate approved a modified version of the Newlands plan, a commission limited to powers of investigation as an amendment to section three of the House Rivers and Harbors bill. In reporting its views on this measure to the conference committee, the Corps took the opportunity to criticize coordinated water administration. In section two, that bill provided that the Engineer Department prepare a national plan for water transportation development. In its memorandum, the Corps recommended that the work which section three entrusted to a commission composed of representatives from all federal water agencies be delegated to the Corps, which alone would report on navigation and "so far as related thereto" on the many other questions which the Newlands commission would investigate.[18] Under this arrangement the Corps could define for itself the functions "related" to navigation, and therefore could stifle the type of investigation contemplated by the Newlands amendment.

The Corps of Engineers also became deeply involved in the public debate over the Newlands bill. Officially and privately, members of the Corps argued that forest cover did not affect stream flow, that reservoirs could not control floods, and that a strong levee system provided the only practical solution to the flood problem. These arguments served as ammunition for those who advocated the levee

[16] Stimson to Newlands, Jan. 27, 1913, in 62nd Congress, 3rd Session, *Sen. Rept. 1339*, 34–35.

[17] Stimson to Newlands, Jan. 22, 1913 (RG #77, #83879/17).

[18] Memorandum in the files of the Corps of Engineers, undated (RG #77, #83879).

program and opposed the Newlands bill. Proponents of multiple-purpose planning, on the other hand, utilized authoritative statements by the Forest Service and the Geological Survey to support their views. The conflict over water policy, therefore, became, in part, a controversy over hydrological theories, and a contest between federal agencies for public support and prestige. The Corps of Engineers eagerly entered this public debate to protect its strategic position in the field of federal water policy.

When Marshall O. Leighton of the Geological Survey in 1907 first presented to the Inland Waterways Commission his proposal to regulate stream flow by means of reservoirs, Brigadier-General Mackenzie, Chief of the Corps, took issue with him.[19] The following spring Leighton presented a summary of his plan in *Engineering News*.[20] A Captain in the Corps, in an issue of the magazine a few weeks later, criticized the Leighton plan.[21] There appeared in *Engineering News*, in rapid order, a reply from Leighton, a second critique by a Major, and another reply from Leighton.[22] At this point the controversy moved into the professional societies, where a Colonel in the Corps took up the attack. In September 1908, in the *Proceedings* of the American Society of Civil Engineers, Colonel Hiram M. Chittenden of the Corps presented a lengthy and vigorous attack on the reservoir and forest cover theories of flood control, under the title, "Forests and reservoirs in their relation to stream flow with particular reference to navigable rivers."[23] At the November meeting of the Society representatives of both sides of the question presented papers. Although hesitating to support the Leighton viewpoint in full, most of the civil engineers at that session considered the reservoir proposal in particular worth examining and

[19] IWC, *Minutes*, Sept. 25, 1907 (GP #2136).

[20] *Engineering News* (May 7, 1908), 59, 498–504.

[21] *Ibid.* (June 11, 1908), 59, 621–624, 624–625.

[22] *Ibid.* (Oct. 22, 1908), 60, 376–380; (Nov. 5, 1908), 60, 504–507.

[23] Hiram M. Chittenden, "Forests and Reservoirs in their relation to stream flow with particular reference to navigable rivers," *Proceedings, American Society of Civil Engineers* (Sept. 1908), 34, 924–997.

were critical of the reluctance of the Corps to explore the idea more thoroughly.[24]

The controversy initiated here continued in the engineering world for many years. From time to time professional journals printed fresh opinions on both sides. In 1910 and 1911 the debate spread from the engineering societies to the general public as Congress considered the Appalachian National Forest bill. The Senate Judiciary Committee had rendered an opinion that the federal government could not constitutionally purchase land for an Eastern national forest unless it would help to improve stream navigation. The controversy over the effect of forests on stream flow, therefore, became a crucial feature of the fight over the Appalachian measure.[25] The conflicting testimony at congressional hearings added little new to the debate. Representatives of the Forest Service and the Geological Survey argued that forest cover did affect stream flow, while the Corps of Engineers and Professor J. Willis Moore, Chief of the Weather Bureau, disputed this testimony.[26]

This debate reached an even higher pitch after the Mississippi and Ohio River floods in 1912 and 1913. Those disasters aroused great interest in a federal flood control program, and intensified the argument over whether Congress should push the levee or the forest-reservoir plan. Popular magazines of nation-wide circulation now carried symposia on the techniques of flood control, the press took sides, and the Corps entered the public debate.[27] In the summer

[24] Discussion of the Chittenden paper in *Transactions, American Society of Civil Engineers* (Mar. 1909), 63, 245–318.

[25] Wells, "Personal History" (GP #1871).

[26] In 1907 the Chief of the Corps had requested private opinions from his divisional and district officers on the effect of forest cover on stream flow (Mackenzie to officers of the Corps, Mar. 29, 1907, NA, War, RG #77, #62743/2). Although these reports, save for a few exceptions, presented little hydrological data on the question, each criticized the claims of the forest enthusiasts (See RG #77, #62743).

[27] *New Orleans Picayune*, July 9, 1912 (GHM, Personal and Biographical, 1); *New Orleans Times-Democrat*, July 9, 1912 (GHM, Personal and Biographical, 1); "Symposium on the Flood's Lesson," *American Review of Reviews* (June 1913), XLVII, 692–704; "Control vs. Prevention," *Cassier's Monthly* (July 1913), 44, 32–46.

of 1913, the Chief of the Corps wrote the President of the Mississippi River Levee Association, "There is only one way to protect the Mississippi Valley from floods, and that is by an adequate system of well-designed levees."[28] The Association widely circulated this statement to vindicate its point of view. Colonel Clifford McD. Townsend, president of the Mississippi River Commission, was the most vocal member of the Corps. Defending his Commission's levee program, Townsend spoke at meetings of engineering societies, drainage organizations, and waterway associations. In 1913 the Senate published his speeches as a public document. Officials of the Corps in Washington became apprehensive over Townsend's activities; when the newspapers announced that he would address a pro-levee political rally in New Orleans, his superiors ordered him not to appear.[29]

After 1913 the Corps of Engineers confined its opposition to multiple-purpose water development to the field of administrative action. In 1913 President Wilson appointed a special Inter-Departmental Committee to consider the Newlands measure, composed of two representatives each from the Departments of Agriculture, Commerce, Interior, and War. In the proceedings of this Committee the Corps clearly revealed its antagonism toward other federal water development agencies. The Committee's final report in April 1914 was signed by all its members except those from the War Department. In their minority report, the War Department representatives maintained that the relations of the Corps "with other Departments and bureaus [were] such that no lack of harmony and of hearty cooperation [was] known. . . . The War Department has cooperated whenever it is practicable to do so."[30] The Committee

[28] Brigadier-General Dan C. Kingman to John A. Fox, Jan. 31, 1914, in Mississippi Valley Levee Association, *A National Duty* (np, nd), 62, copy in RG #77, #83879/152).

[29] 63rd Congress, 1st Session, *Sen. Doc. 204, Mississippi River Floods*; Walter Parker to Newlands, Nov. 27, 1913 (FGN, Waterways-River Regulation, Correspondence, 1913, 7).

[30] "Report of the War Department Representatives in Respect to the Newlands Bill, Supplemental to the report of the Inter-Departmental Committee of January 10, 1914," April 13, 1914 (RG #77, #83879/142).

did not publish either of its reports, but the minority view of the Corps, in mimeographed form, reached Mississippi Valley pro-levee Senators, and thereby all of Congress.[31] Newspaper accounts noted particularly that the Committee could not agree and that the War Department would submit a minority report.[32]

Several years later Secretary of War Newton D. Baker described the attitude of the Corps and the problem it created:

I understand when I came [into the cabinet] that a difference of opinion had developed between Mr. Garrison and other members of the cabinet, chiefly, so far as I could analyze it, from the fact that the Secretary of War felt that the engineer department, which had for a great many years been giving expert and more or less exclusive attention to the subject of navigable rivers, was so well equipped to continue that subject that it would be a pity to divide its responsibility and authority by co-ordinating it with any other branch of the government.[33]

Baker gave strong support to the Newlands bill. He worked with the Senator to devise an acceptable measure, reported favorably on it to Congress, and spoke for it in cabinet meetings.[34] Under such a Secretary, the Corps felt stymied. A letter to the Chief of the Corps in August 1916 from Colonel Harry Taylor, a member of the former Inter-Departmental Committee, revealed the predicament of the Engineers. The Chief had forwarded Baker's correspondence to Taylor, and Taylor replied:

I infer from the correspondence between the Secretary of War and Senator Newlands that the Secretary is a convert to the Newlands' ideas, and if he is it seems to me that the only thing which we can do now is to endeavor to have proper representation upon the Board . . . which

[31] Frederick H. Newell to Newlands, June 8, 1914 (FGN, Waterways-River Regulation, 1912–1916, 3); War Department memorandum to Sen. Ransdell, with copy of the report, April 16, 1914 (RG #77, #83879).

[32] *New Orleans Times-Democrat*, Apr. 10, 1914 (GHM, Clippings, v. 17).

[33] *Washington Star*, Dec. 23, 1916 (GP #1665).

[34] Baker to Newlands, Aug. 1, 1916; Baker to Newlands, Aug. 11, 1916 (both in FGN, Waterways, 1916, 2); Baker to Newlands, Dec. 21, 1916 (NA, RG #77, #83879/230); Baker to Committee on Commerce of the Senate, Jan. 5, 1917 (RG #77, #83879/233).

will necessarily be formed to carry out the work of the . . . National Waterways Commission. . . . I do not see how you can do anything in opposition to the bill. What you can do, if the Engineer Corps is to obtain proper recognition, if the bill should pass seems to me to be something to be determined by developments. . . . It rather looks as if it had gotten out of our hands for the present.[35]

The Chief of the Corps, Major-General William M. Black, did not despair. Although Congress approved the Newlands commission, limited to powers of investigation, as an amendment to the Rivers and Harbors Act of 1917, the Corps successfully blocked its appointment by the President. Preoccupied with the War, Wilson could not attend to the commission; neither could Secretary of War Baker. Consequently, the President forwarded all correspondence on the subject to General Black who deposited it in his files, "pending such time as the President may desire to take action in the matter." To the Secretary of War the General reported, "It is my opinion that the situation is not such as to warrant the organization of the commission and the active prosecution of duties entrusted to it at this time. . . . The consideration of this whole subject should be postponed."[36] As George Maxwell argued to the War Department, the Chief of the Corps was pressing Wilson not to execute a law passed by Congress.[37] Following Black's advice, the President did not appoint the commission.

The Root of Corps Opposition

The hostility of the Corps of Engineers to the Newlands bill sprang from three sources: its interest in navigation alone, its willingness to defer to Congress, and its desire for complete autonomy in administration. While the Forest Service, the Geological Survey, and the Bureau of Reclamation considered water as a single resource with many possible functions, the Corps limited its concern to navi-

[35] Col. H. Taylor to General Black, Aug. 28, 1916 (RG #77, #83879/247).

[36] Major-General William M. Black, "Memorandum for the Secretary of War," July 2, 1918 (GHM, A 7).

[37] Maxwell to Dr. F. P. Keppel, Assistant Secretary of War, July 17, 1918 (GHM, A 7).

gation. When measures pending in Congress were submitted to the Chief of the Corps for his comments, he frequently replied. "The provisions of the act amply safeguard the interests of navigation, and, so far as those interests are concerned, I see no objections to approval by the President."[38] Congress had never given the Corps authority to deal with matters other than navigation, and many, including its officers, argued that Congress could not constitutionally enact such a measure. The Corps's exclusive interest in navigation, however, stemmed even more fundamentally from its failure to grasp the possibilities of the concept of multiple-purpose river development.

Officers of the Engineer Department invariably spoke of navigation as a "primary" feature of river development and other uses of water as "secondary" or "related to" navigation. A district engineer of the Corps once declared, "The development of water power should . . . be made always secondary to navigation interests."[39] In 1907 Brigadier-General Mackenzie had told the Inland Waterways Commission that soil erosion was a soil, not a water problem.[40] In his report dissenting from the findings of the Commission, he declared, "I am not fully in accord . . . with the thought that all the related subjects mentioned . . . are as clearly and necessarily associated with the subject of channel improvement and interstate commerce as is assumed in the report. . . ."[41] And the War Department's minority report from the Inter-Departmental Committee on the Newlands measure declared that the Corps had investigated the bill to determine the feasibility of irrigation, flood control, and power development "in relation to navigation improvement."[42]

In the Rivers and Harbors Act of July 25, 1910, Congress directed

[38] E. W. Burr, acting Chief of Engineers, to Secretary of War, June 22, 1910 (RG #77, #49678/28).

[39] Major Clement A. F. Flagler to Mackenzie, Apr. 8, 1907 (RG #77, #63743).

[40] IWC, *Minutes*, Oct. 7, 1907 (GP #2136).

[41] IWC, *Report*, 25–27.

[42] "Report of the War Department Representatives in Respect to the Newlands Bill, Supplemental to the Report of the Inter-Departmental Committee of January 10, 1914," April 13, 1914 (RG #77, #83879/142).

the Corps to investigate the possibility of developing water power jointly with navigation to help finance river improvement. Here the Engineers had a wonderful opportunity to broaden their conception of river development, but their views remained narrow. In reports under this Act they considered water power not as an aid to improve streams which might not warrant development for navigation alone, but as a by-product to be considered after a navigation project had been approved on its own merits.[43] Colonel Townsend, of the Mississippi River Commission, best stated the implications of this narrow attitude. "While there are exceptions," he declared, "it is usually advisable to limit the development of a stream to the particular purpose for which it is best adapted.... What is most needed [referring to the Newlands proposal for a waterways commission] is a Board to differentiate instead of coordinate."[44]

In self-defense, however, the Corps of Engineers entered the field of flood protection. Local pressure for federal aid to prevent flood damage, especially in the lower Mississippi Valley, persisted throughout the middle of the nineteenth century.[45] In 1879 Congress finally created the Mississippi River Commission, and in 1893 the California Debris Commission, to tackle flood problems on the Mississippi and the Sacramento-San Joaquin river systems. Since many feared that the courts would frown on a federal flood program as unconstitutional, Congress undertook these tasks under the guise of improving navigation. And, since the Corps hesitated to become involved formally in flood control, the lawmakers established commissions of army engineers independent from the Corps to execute the plans.

By 1913 interest in a broader federal flood control program had risen until it appeared that if the Corps did not fall in line, Congress would grant other federal agencies the task of administering a new

[43] 62nd Congress, 3rd Session, *House Doc. 1410, War Department Report on the New River,* for example; Philip P. Wells, "Memorandum for Kent with respect to the Adamson Bill," July 13, 1914 (GP #1665).

[44] Col. C. McD. Townsend, "Newlands River Regulation Bill" (RG #77, #83879/182).

[45] For Mississippi flood control see Arthur DeWitt Frank, *The Development of the Federal Program of Flood Control on the Mississippi River* (New York, 1930), 134–156.

measure. The disastrous Ohio Valley flood of that spring brought the problem to a head. For a number of years a privately organized Pittsburgh Flood Commission, after completing an exhaustive three-year investigation of Pittsburgh flood problems, had sought federal funds to carry out its recommendations. These plans emphasized both reservoir and forest cover flood control.[46] Since the Corps opposed these methods, its officers knew that Congress would entrust such a program to a more sympathetic federal agency. Therefore, on their own initiative, and without a congressional directive, local officials of the Corps at Pittsburgh began to collect data to counteract the report of the Pittsburgh Commission.[47] In this struggle for control of the program, the Corps won out. In 1913 Congress appointed a commission of army engineers to investigate the Ohio Valley flood problem.[48] From that time on the Corps of Engineers willingly undertook flood control tasks, although, in doing so, it did not accept the concept of multiple-purpose development and confined its recommendations to levee construction.

The Corps of Engineers opposed the Newlands measure also because it did not wish to antagonize Congress. Its officers realized that the lawmakers preferred to select public works projects through logrolling rather than to bestow the task upon an executive commission; if the Corps became too independent and tried to force its views upon Congress, the lawmakers might transfer jurisdiction over navigation to other federal agencies. The Geological Survey and the Bureau of Reclamation derived their political support largely from the executive branch of the government. The very existence of these agencies, the Corps feared, threatened the predominant position which the Engineer Department had long held in the field of river development. To counteract the influence of these new

[46] Pittsburgh Flood Commission, *Report of Flood Commission of Pittsburgh, Pa., 1912* (Pittsburgh, 1912).

[47] George M. Lehman to Lt. Col. H. C. Newcomer, Nov. 3, 1911 (RG #77, #72788/779).

[48] *Chicago Inter-Ocean*, Apr. 13, 1914 (GHM, Clippings, v. 18); Louisville *Courier-Journal*, Apr. 22, 1914 (GHM, Clippings, v. 18); *Chicago Tribune*, Apr. 27, 1914 (GHM, Clippings, v. 18).

agencies, and to mend its political fences, the Corps was forced to cultivate Congress.

The Corps had already learned its lesson when it had tried to discourage the growing congressional enthusiasm for water transportation. Officers of the Corps agreed with Representative Burton, Chairman of the Rivers and Harbors Committee, that the public overestimated the value of river transport, and that many streams could not feasibly be made navigable.[49] The Corps and Burton worked together closely to thwart the new demands, but in vain. Dissident waterway groups took possession of the Rivers and Harbors Committee in 1910 amid threats to transfer to a civilian agency the navigation activities of the Corps. The Engineers soon yielded. For example, in 1903, on the advice of the Corps, Congress discontinued the Missouri River Commission, which it had organized earlier in 1895 to supervise the development of navigation on that river. But, when local waterway enthusiasts persisted, Congress in 1912 authorized a six-foot navigable channel from Kansas City to St. Louis. After surveying the project, the Corps was prepared to turn in an adverse report. The Valley arose in protest, and, yielding to the pressure, the Engineers approved the project.[50] Burton, now a Senator, looking on with dismay as his old allies no longer resisted, publicly accused them of weakening and submitting to congressional logrolling.[51]

Such experiences rendered the Corps very sensitive to the wishes of Congress. Although it would have preferred to develop waterways more systematically, the Corps hesitated to take the initiative in recommending that Congress adopt a new approach. The Chief of the Corps did so once, but nothing came of it. In his Report for 1911, the Chief suggested that the Engineers, if Congress desired,

[49] "Report of the War Department Representatives in Respect to the Newlands Bill, Supplemental to the Report of the Inter-Departmental Committee of January 10, 1914," April 13, 1914 (NA, War, RG #77, #83879/142); memorandum on the Newlands bill (1907), unsigned, but on letterhead of the Chief of Engineers (RG #77, #62743).

[50] Pross, "Rivers and Harbors," 175–176.

[51] 62nd Congress, 2nd Session, *Congressional Record*, 4796 ff.

could draw up plans for a national water transportation system.[52] The House Rivers and Harbors Committee responded to this feeler only to the extent of authorizing the Corps to make inquiries regarding terminals and transfer facilities.[53] Taking up the Chief's suggestion, Senator Newlands persuaded the Senate Committee on Commerce to amend the Rivers and Harbors bill of 1912 by authorizing the Corps to prepare the plan. When the conference committee refused to accept this suggestion, the plan was dropped.[54]

By this time Congress had completely cowed the Corps into submission. In 1919 Henry L. Stimson, who had been Secretary of War from 1911 to 1913, described the situation:

When I was Secretary of War I found this situation, and I found that the reports of the Chief of Engineers which came to me were not "Is this an improvement which should be made in view of our particular funds this year—our particular budget this year—and in view of all the improvements in the United States taken at the same?" but simply and solely "Is this an improvement of a waterway which should be made?" And the Chief of Engineers said he was directed by Congress to report in that way, and this was the way he was going to interpret that, not in comparison with other projects, but simply whether in the millennium it would be a good thing for the country to have the waterway improved. When I said, "That does not suit me at all. You come in here with a lot of propositions which you have approved, and you want me to approve, to improve the navigation of such and such a river and such and such a creek and such and such a harbor. I want to know how does that compare with the situation of the whole?" He said, "I have nothing to do with that. I cannot have anything to do with it. Congress will not listen to me on that. They reserve the judgment to do that themselves."[55]

While Stimson blamed the Corps for its inaction, Senator Newlands

[52] *Annual Report of the Chief of Engineers, 1911*, 35; Newlands to Maxwell, Mar. 4, 1912 (FGN, Waterways-River Regulation, Correspondence, 1912, 1).

[53] *Ibid.*

[54] Newlands to Maxwell, May 1, 1912 (FGN, Waterways-River Regulation, Correspondence, 1912, 2).

[55] 66th Congress, 1st Session, House Select Committee on the Budget, *Hearings on the Establishment of a National Budget System*, 641.

was inclined to find the cause for the trouble in Congress. He declared:

The Engineer Corps of the Army has been put in a strait-jacket by Congress itself. In one of my speeches I made numerous quotations from river and harbor bills absolutely forbidding the Engineer Corps to make any suggestions outside of the matter particularly submitted to them; and the very limited response to Gen. Bixby's suggestion made by the River and Harbor Committee in the House bill indicates their willingness to see that money is spent upon the rivers, but their indisposition to have it spent in a really effective way.[56]

Newlands, in part, wrote correctly. One can trace much of the attitude of the Corps to Congress itself. Yet, the Corps's narrow conceptions of river development and of administrative responsibility sprang from its own limitations as much as from congressional pressures. Other federal departments continued to speak out; the Corps did not. When its suggestions for changes elicited no favorable response, the Engineers quickly ceased to give advice. While others spoke of plans that were most desirable from a technical point of view, the Corps emphasized those that were politically feasible. Officials of the Corps realized that congressional attitudes thwarted an effective water transportation system. They also knew full well that if they insisted on a more constructive approach to waterway problems, Congress might limit seriously their administrative independence.

Such restricting of their independence, officials of the Corps feared most of all. Desire for complete authority and autonomy both cemented their attachment to Congress and heightened their hostility to the Newlands bill. In 1913, for example, the Corps approved a sweeping investigation of water problems if it, rather than an independent commission, were to undertake the task.[57] During the early years of the twentieth century, the Corps faced the possibility

[56] Newlands to Maxwell, Mar. 4, 1912 (FGN, Waterways-River Regulation, Correspondence, 1912, 1).

[57] Memorandum in the files of the Corps of Engineers (RG #77, #83879).

of coordination with municipal, state, and private enterprise, as well as federal agencies. In each case the Engineers drew back and chose to go it alone. They feared any limitation on their own activities which another agency, private or public, might impose.

The attitude of the Engineers toward water power development illustrates this anxiety. Proposals frequently arose for joint public-private multiple-purpose water power projects in which the federal government would construct and operate the dam, and private enterprise the hydroelectric installation. The Corps frowned on this arrangement. Brigadier-General Mackenzie, Chief of the Corps, emphasized his opposition as early as 1905:

> Locks and dams are built and operated for the purpose of facilitating navigation and commerce, and nothing should be permitted that would tend to impair their usefulness, or interfere with their operation for this purpose. Partnerships or quasi partnerships between the Government and private persons or corporations have not been generally favored in the past, as experience has shown that they are apt to be attended by many annoying complications.[58]

Brigadier-General Hiram M. Chittenden explained more fully the reason for the Corps's attitude on this question.

> If private interests were to build the plant, they would acquire vested rights which would always stand in the way of future control and lead to complications if it should become necessary to terminate the arrangement. With the plant in the possession of the Government and the users standing simply in the relation of lessees for a limited period, without great initial expense on their part, and with freedom on the part of the Government to control the arrangement without the complication of private ownership, the whole plan would stand on a simple, practical business basis.[59]

The Corps also became increasingly fearful that construction of privately owned dams on navigable streams would interfere with its

[58] Brig.-Gen. Alexander Mackenzie to Secretary of War, Jan. 16, 1905, in War Department Circular No. 14 (GP #1665).

[59] Hiram M. Chittenden, "Forests and reservoirs in their relation to stream flow," 34, 989–990.

operations. Before 1910, the Corps had been content to retain power to prevent hydroelectric installations from interfering with navigation. In granting permission for private corporations to construct dams in navigable streams, for example, Congress explicitly required the owners to abide by any regulations the Secretary of War might prescribe to safeguard navigation.[60] Under these acts private corporations retained little authority. The Secretary of War could maintain pool levels for navigation and restrict their use for water power, and private parties, under such conditions, could enter no claims against the government. To ensure the supremacy of navigation, these statutes contained such clauses as: "that the use of such power shall in no instance impede or hinder navigation," "no claim shall be made against the United States for any failure of water power, resulting from any cause whatsoever," "that the right to alter, amend, and repeal this Act, and the right to require the alteration or removal of the structure authorized without any liability on the part of the United States, are hereby expressly reserved."[61]

As the construction of private dams increased in the early years of the twentieth century, these sweeping safeguards no longer satisfied the Corps. Soon they argued that the federal government alone should develop water power on navigable streams. In November 1911, Colonel William M. Black, future Chief of the Corps, but then Division Engineer at New York City, warned:

> [In erecting such dams] the creation of power is the primary object, the needs of navigation the secondary. In times of small volume of stream discharge there is the constant tendency to use the waters for power purposes, to such a degree that the pool depths are lowered unduly and navigation impeded or stopped. The protection of navigation is a duty of the United States held in trust for the entire nation and this duty cannot be delegated. The immediate interests of a community frequently conflict with those of the whole nation.[62]

In 1905 the Corps had advocated that private power installations in

[60] For a compilation of all such statutes prior to 1908 see IWC, *Report*, 597–696. [61] 33 *Stat.* 1132; Act of June 4, 1906, 34 *Stat.* 211; 34 *Stat.* 265.

[62] Col. William M. Black to Chief of Engineers, Nov. 16, 1911 (RG #77, #73788/791).

navigable streams be subject to state law with only limited federal restrictions.[63] But Colonel Black now argued:

[The] use of navigable waters of the United States for power purposes by a state, or under state authority, is subject to grave dangers . . . it would seem that the entire control of the navigable waterways of the United States must lie in the United States and that the natural line of division between National and State control of water power must be determined upon that basis.[64]

Another member of the Corps spoke in more explicit terms:

Since the nation must exercise control over navigable waters of the United States in the interest of navigation, it is thought that the inexpediency of divided control places the line of demarcation between State and National control of water power, exactly where it now rests with regard to the improvement of navigation. In other words, water power on navigable waters of the United States should be developed, owned and utilized by the United States and that on other navigable waters, by the States within which they lie.[65]

The Corps increasingly feared the effects of private construction and operation of reservoirs on tributaries as well as on the main stems of navigable streams. In opposing the Newlands bill, army engineers officially denied that upstream reservoirs could affect navigation appreciably, but privately they expressed great concern and argued that the Corps should control them. Colonel Black observed that storage reservoirs were operated in conjunction with hydroelectric installations, and could therefore greatly affect navigation. When a reservoir is "built primarily with the object of power production," he argued, "it will be designed and operated as is deemed most advantageous for this object, and not for the benefit of navigation. It would seem the United States would be compelled to exercise a control over such reservoir systems when built."[66]

[63] Brig.-Gen. Alexander Mackenzie to Secretary of War, Jan. 16, 1905, in War Department Circular #16 (GP #1665).

[64] Col. Black to Mackenzie, Nov. 16, 1911 (RG #77, #73788/791).

[65] Lt. R. D. Black to Col. W. M. Black, Nov. 20, 1911 (RG #77, #72788/808).

[66] Col. W. M. Black to Mackenzie, Nov. 16, 1911 (RG #77, #72788/791).

The Corps's investigation of the 1913 floods in the upper Ohio Valley provided clinching evidence to its officers that the federal government should control nonnavigable as well as navigable streams. Private corporations and municipalities, the Engineers argued, contributed heavily to flood damage by constantly encroaching on stream beds in constructing bridges, piers, and wharves. To solve the problem, so the Corps and Secretary of War Garrison reported, Congress should bring nonnavigable streams under the jurisdiction of the War Department, and require that the Chief of the Corps approve the erection of all structures which might augment flood damage. Heretofore, the Corps had argued frequently that the states alone controlled these tributaries, but the trend of affairs in both hydroelectric development and flood control now compelled it to expand greatly its conception of the proper sphere of federal jurisdiction. Flood control, as well as navigation, the Corps now realized, required a single federal authority, rather than a combined jurisdiction which invited conflicts.[67]

[67] Louisville *Courier-Journal*, Apr. 22, 1914 (GHM, Clippings, v. 18); Louisville *Courier-Journal*, Oct. 2, 1913 (GHM, Clippings, v. 16); Theodore E. Burton to Lindley M. Garrison, nd. (RG #77, #83879/173); Garrison to Burton, June 25, 1914 (RG #77, #83879/178).

Chapter XI

Congress Rejects Coordinated Development

The Newlands measure threatened to limit the authority of Congress over water development as clearly as it would restrict the independence of the Corps of Engineers; the lawmakers could not ignore its implications. Many organizations supported Senator Newlands, as the best hope of securing congressional commitment for their projects; they did not worry, for the moment, about restrictions which might be imposed by an independent commission which they could not control. Alarmed at the increasing popularity of the Newlands measure, the lawmakers gradually agreed to enter new fields of water development, but in such a way that Congress itself, rather than an independent commission, could control the choice of projects and their annual appropriations. Although Congress did not immediately grant all of the demands presented by organized interest groups, its initial gesture of support sufficed to lure most of them from the broader Newlands measure to the pattern which Congress preferred. This loss of political strength gradually weakened the Newlands bill until by 1917 it rested primarily upon the personal power and prestige of the Senator himself.

The Changing Attitudes of Interest Groups

Initial enthusiasm for the Newlands Commission came from river transportation groups which bitterly criticized the House Rivers and Harbors Committee and the Army Corps of Engineers for refusing to favor their demands. In 1907 they looked upon the Inland Waterways Commission as their major hope, for its members seemed more inclined to approve of the new interest in large-scale waterway development. In the spring of that year, Representative Burton of Ohio, the major congressional opponent of the new waterway enthusiasts, had tried to prevent the creation of the Inland Waterways Commission by proposing a similar investigating body to be dominated by Congress. Of its nine members, the Vice-President would select three, the Speaker of the House three, and the President three men "versed in transportation questions." The proposed group would advise Congress about needed river and harbor improvements for navigation.[1] Although Burton's move failed, Roosevelt, unable to ignore his influence and ability, appointed him chairman of the Inland Waterways Commission.

When the Inland Waterways Commission failed to recommend the specific improvements it desired, the Rivers and Harbors Congress in 1908 demanded that another waterway commission be appointed.[2] Four waterway leaders, joined by flood control representatives, introduced this proposal in the House in January 1909. To circumvent Burton's influence, the bill was referred to the Committee on Foreign and Interstate Commerce instead of to Burton's Rivers and Harbors body. But the representative from Ohio outsmarted his opponents by proposing, successfully, a similar commission in an amendment to the Rivers and Harbors Act of 1909. The Rivers and Harbors Committee and Burton, therefore, dominated the new group's membership.[3] The six Representatives and five Senators on this National Waterways Commission exhaustively studied flood and transportation problems in Europe as well as in the United States. Their report outlined a conservative policy. It accepted Burton's theory that, although waterways helped to lower transportation rates, such changes should come through federal and state rate regulation, rather than through waterway competition. It approved reservoir stream control in theory, but, because of its expense, not in practice. It rejected an expanded federal flood control

[1] D. M. Bone to Roosevelt, March 6, 1907 (NA, War, RG #77, #63743/2).
[2] 60th Congress, 2nd Session, *Congressional Record*, 180. [3] *Ibid.*, 2821.

plan. Following Burton's lead, the National Waterways Commission tried to dampen the enthusiasm for extended river transportation development.[4]

Although Burton succeeded temporarily, he could not continue to stem the tide. Alarmed at the original popularity of the Inland Waterways Commission and the Newlands measure, both the Corps of Engineers and the Committee on Rivers and Harbors abandoned him and responded to the new demands.[5] When Burton moved to the Senate in 1908, DeAlva S. Alexander of New York, far more friendly to waterway development, became the new Committee chairman. At the opening of the 61st Congress, moreover, fully one-third of the Committee's members were new appointees who shared this viewpoint. In presenting his first measure to the House in 1910, Chairman Alexander announced that the Committee had adopted a new policy. Under Burton, he explained, infrequent bills had provided appropriations for only a few of the larger projects. Henceforth, yearly bills would give the Committee an opportunity to consider more projects.[6] This innovation in Committee policy enabled most of the waterway associations to obtain approval for their measures through established channels. Consequently, they withdrew their support from the Newlands bill and worked through the Rivers and Harbors Committee and the Corps of Engineers.

The Lakes-to-the-Gulf Waterway Association did not share in this success. In the Rivers and Harbors Act of 1910 that group obtained an appropriation of $1,000,000 for a special Army Engineer commission, which the president would appoint, to investigate their project. President Taft, however, angered the Mississippi Valley by selecting the members of the commission only after much delay, and by choosing men already committed against the proposal.[7] In

[4] 62nd Congress, 2nd Session, *Sen. Doc. 469*, 1–62.

[5] Even Burton stated publicly that he would favor $50,000,000 a year for river and harbor improvements. See *Waterways Journal* (Jan. 25, 1908), 20, 5.

[6] Pross, "Rivers and Harbors," 163 ff.

[7] *Proceedings, Lakes-to-the-Gulf Deep Waterway Association, 1910*, 154–164.

1912 the Association secured a hearing before the Rivers and Harbors Committee, but it made little progress.[8] Moreover, in 1915 its major foe, Senator Burton, persuaded Congress to repeal the 1910 provision for a long-term investigation of the Lakes-to-the-Gulf plan. As a result of these failures, the Newlands bill still looked attractive to the Mississippi Valley. Robert R. McCormick, a prominent Illinois conservationist, expressed the Valley's viewpoint when in 1912 he declared,

> We see in the reclamation laws a vision of the ideal when we compare them with the pitiable log-rolling, pork-barrel rivers and harbors bills that have disgraced this nation year in and year out for generations. . . . We have to look forward to a comprehensive, businesslike conservation and development of all our resources. . . . We have gone far enough where we will welcome any kind of a reasonably successful federal law either by the appointment of a commission or otherwise to regulate to its best use all the water which falls upon the surface of our land, bearing in mind irrigation, navigation, drainage, and all the uses to which this water can be put; and the prevention of all the harm which it can do if misused.[9]

Leaders of the Lakes-to-the-Gulf movement lobbied for the Newlands bill in Congress, pressed the President to select a new Chief of the Corps more friendly to their views, and defended the reservoir and upstream watershed control theories.[10] By the 1920's, the Lakes-to-the-Gulf Association remained almost alone among active waterway groups in support of multiple-purpose river development.

Senator Newlands received added support, for a brief time, from those who advocated a federally financed drainage program. But, when it appeared that Congress would look with greater favor upon a bill limited specifically to their one proposal, the drainage organizations quickly lost interest in the Newlands measure. As early as

[8] William K. Kavanaugh to McGee, Feb. 22, 1912 (WJM, box 1909–1916).

[9] *Proceedings, National Irrigation Congress*, 1912, 2–3. McCormick was chairman of the board of trustees of the Chicago Sanitary District, president of the Illinois Conservation Association, and president of the Illinois Rivers and Lakes Commission.

[10] William K. Kavanaugh to Newlands, Sept. 22, 1913 (FGN, Waterways-River Regulation, Correspondence, 1913, 5).

1901 Congress provided for technical aid to drainage projects by creating in the Department of Agriculture an Office of Drainage Investigation.[11] In 1906 Senator Hansbrough of North Dakota tried, without success, to establish in the federal treasury a North Dakota drainage fund, to consist of receipts from the sale of public lands in that state, and to finance drainage projects there. Other states with large areas of swampland, such as South Carolina, Georgia, Florida, Wisconsin, and Missouri, developed interest in a similar federal program. In November 1906 these groups held a National Drainage Congress in Oklahoma City to form a nation-wide lobbying organization, and Representative Steenerson of Minnesota and Senator Flint of California introduced measures in Congress to carry out their program.[12]

The Drainage Congress soon collapsed, however, and was replaced by a more limited, but more effective organization, spearheaded by businessmen interested in reclaiming swamplands along the Mississippi River. Railroads in the Valley helped to promote this venture. In the summer of 1911, for example, the Illinois Central Railroad sponsored a "Reclamation Special," which toured southern Illinois, Kentucky, Tennessee, Louisiana, and Arkansas, booming land drainage.[13] The Louisiana Development League, a group of real estate promoters, in 1911 tried to convert the National Irrigation Congress into a national drainage association to lobby for federal aid.[14] Chicagoans (the Congress met in the Windy City), who visualized the possibility of combining drainage and deep waterway enthusiasts in one political effort, cooperated, but Western members, who feared that the venture would endanger the irrigation program, resisted the move, and continued in control. The Louisiana leaders then organized a separate National Drainage Association, which met in New Orleans in the spring of 1912.[15]

[11] *Engineering News* (July 24, 1902), 48, 64.

[12] *Forestry and Irrigation* (Jan. 1907), 13, 35; flier publicizing the first annual meeting in GP #2137. [13] *Irrigation Age* (July 1911), 26, 974, 979.

[14] A. B. Graves to Board of Directors and members of Louisiana Development League, Mar. 12, 1912 (GHM, Invoices).

[15] *Proceedings, National Irrigation Congress, 1911*, 32–34, 66–83, 124–125.

Senator Newlands tried to attract this new movement to his measure by stressing that his proposal included funds for drainage as well as irrigation, flood control, navigation, and water power.[16] He spoke at the second meeting of the National Drainage Congress at Baltimore in 1907 on the topic, "National Drainage and its Effect on Inland Waterways." In 1912 he persuaded the National Drainage Association to make a determined, even though brief, effort to push the Newlands bill. At first the multiple-purpose measure looked extremely promising to the Drainage Association. Its president, Edmund T. Perkins, a former official of the Bureau of Reclamation, declared the bill to be "all that the drainage enthusiasts could wish for. . . ."[17] But Perkins also knew of the difficulty of obtaining concerted congressional action for a measure so broad. Some leaders of the Association favored, as a more practical venture, a National Drainage Service, similar to the Bureau of Reclamation, limited to one specific task.[18] Congress appeared to prefer this approach. In March 1913, in an interview with Champ Clark, Speaker of the House, Perkins was instructed to come to Congress with something definite, not a "policy," but a "project."[19] The lawmakers, desiring something more specific, had difficulty in tackling a measure as broad as the Newlands proposal.

This advice from Speaker Clark divided the drainage organizations. The Drainage Association, now composed of groups from all sections of the country, switched to a measure limited to drainage alone. During the summer of 1913 the Association prepared a bill to authorize the Secretary of the Interior to undertake a drainage program and to apportion expenses among the landowners, cities and states whom it would benefit. The federal government would

[16] Newlands to D. G. Purse, Jan. 7, 1908; Newlands to C. P. Goodyear, Jan. 17, 1908 (both in FGN, Waterways-River Regulation, Correspondence, 1907–1911, 1).

[17] News release from the National Drainage Association in letter, Philip R. Keller to Newlands, Feb. 16, 1912 (FGN, Waterways-River Regulation, Correspondence, 1912, 1).

[18] Jacob A. Harman to Newlands, Jan. 29, 1912 (FGN, Waterways-River Regulation, Correspondence, 1912, 1).

[19] *Dallas News*, May 24, 1913 (GHM, Clippings, v. 16).

finance only that portion of the total project which pertained to the national welfare. Secretary of the Interior Franklin K. Lane approved the measure, and, by request, Speaker Clark introduced it in the House.[20] Louisiana groups, however, concerned with flood control as well as drainage, split off from the national organization and continued to endorse the Newlands bill. They took no part in the agitation for a measure limited to drainage, and even opposed it as a threat to their interests.[21]

Flood control leaders in the lower Mississippi Valley, seeking federal financial aid, also flirted temporarily with the Newlands bill. Since 1879 the Mississippi River Commission had supervised the long project of bringing the river under control. Though ostensibly engaged in promoting navigation, the Commission concentrated on flood protection, and on building, repairing, and strengthening levees. Local levee boards undertook most of the financial burden of this work. Levee districts, organized under state law, floated bonds to raise funds, taxed property to repay them, and supervised local levee construction. As the financial burden grew, however, taxpayers in the Valley demanded that the federal government assume the cost.[22] The Interstate Mississippi River Levee Improvement Association, formed in 1890, led the fight for federal aid. "The National Government," it argued, "[should] assume and control the construction and maintenance of these levees as an aid to River Improvement and for the protection of the Valley from floods. . . ."[23]

[20] *Chicago Inter-Ocean*, Aug. 23, 1913 (GHM, Clippings, v. 16); *New York Times*, Sept. 13, 1913 (GHM, Clippings, v. 16).

[21] Maxwell to Newlands, Mar. 30, 1912 (FGN, Waterway-River Regulation, Correspondence, 1912, 1); Edmund Perkins to Newlands, Mar. 24, 1913 (FGN, Waterways-River Regulation, Correspondence, 1913, 2); Pinchot to John F. Bass, Feb. 14, 1914 (GP #1827).

[22] Frank, *Flood Control*, 134–156.

[23] See "Call for Levee Convention," signed by Charles Scott, president of the Interstate Mississippi River Improvement and Levee Association (FGN, Inland Waterways, 1907–1910, 2).

This proposal met resistance similar to that encountered by waterway enthusiasts. The Corps of Engineers, which carried out flood control work in the Valley under the direction of the Mississippi River Commission, argued that the federal government should not undertake this new financial responsibility.[24] Representative Theodore E. Burton opposed federal appropriations for flood control as vigorously as he resisted new navigation projects. He hoped that limited and temporary federal aid would suffice to strengthen local levee districts so that they could bear the entire burden. The National Waterways Commission of 1909–1912, which Burton dominated, concluded emphatically that no national interest in flood prevention warranted large federal expenditures. Moreover, many argued that the federal government could not constitutionally undertake flood protection.[25] Doubt over this question had prompted Congress to create the Mississippi River Commission as an instrument to aid navigation when, as everyone knew, it dealt primarily with flood protection.

Disastrous floods in the spring of 1912 and again in 1913 brought the Mississippi River to the attention of the entire country, stirred the Valley to renew the campaign for federal aid, and increased congressional support for the plan. In its report for 1912 the Mississippi River Commission admitted for the first time that its work was devoted primarily to flood protection.[26] Shortly after the spring floods of 1912, levee boards along the middle Mississippi River formed a new group, the Mississippi Valley Levee Association. Its leadership and financial support came primarily from owners of large plantations, mortgage banks, holders of levee district bonds, and railroads and industries whose property the floods had damaged

[24] Testimony of Capt. G. M. Hoffman in MSS of meeting of the Inland Waterways Commission, May 18, 1907 (GP #2136).

[25] Frank, *Flood Control*, 140–143; see also Leahmae Brown, "The Development of National Policy with respect to Water Resources," doctoral dissertation at the University of Illinois (Urbana, 1937), 55.

[26] Frank, *Flood Control*, 143.

severely.[27] These groups based their campaign upon the statement of the Mississippi River Commission that a levee system of sufficient height and thickness would provide complete protection. Senator Joseph E. Ransdell of Louisiana and Representative Benjamin C. Humphreys of Mississippi introduced into Congress a bill authorizing federal construction of the proposed levees at a total cost of $60,000,000.

This program did not please the lower valley. Although levee construction had aided the middle Mississippi area, it had also enhanced the flood danger further downstream by confining the river to an increasingly narrow channel, and by forcing a greater volume of flood water into the lower reaches of the river. To solve this problem the lower valley first proposed to divert part of the river to the ocean through a new channel, and thereby to remove much of the strain from the main stream. The Mississippi River Commission, however, disapproved this outlet system and confined itself to levee construction. Senator Newlands, on the other hand, included the outlet proposal as one of the provisions of his measure. The lower valley repudiated the Commission's engineering advice, opposed the Ransdell-Humphreys plan, and supported the Newlands bill.[28]

As early as the fall of 1911, George Maxwell, Senator Newlands' former partner in the campaign for a federal irrigation program, began to organize agitation in New Orleans for the multiple-purpose development bill. Here he received strong support. Using the technique that had brought him success in the irrigation campaign, Maxwell appealed especially to manufacturing concerns with markets in the lower valley. He brought together businessmen involved

[27] John A. Fox, *A National Duty* (np, nd), an account of the work of the Mississippi Valley Levee Association, written by its secretary, and distributed for publicity purposes. A copy, with Secretary of War Garrison's name printed in gold on the cover, is located in RG #77, #83879/152.

[28] Peter S. Lawton to the editor of the *New Orleans Picayune* in the *New Orleans Picayune*, July 12, 1912 (GHM, Invoices); Frank, *Flood Control*, 112–121; *New Orleans States*, July 14, 1913 (GHM, Clippings, v. 16).

in the development of southern Louisiana swamplands.[29] In search of methods of flood protection other than levees, these promoters found the reservoir and upstream watershed techniques of flood control especially attractive.[30] In 1912 Maxwell organized them into the Louisiana Reclamation Club. This group, in turn, became the nucleus of a nation-wide organization, the National Reclamation Association which spearheaded the fight for the Newlands measure.[31]

Between these two groups, advocates of the Ransdell-Humphreys bill and those who favored the Newlands plan, a bitter controversy raged in the lower and middle Mississippi Valley between 1912 and 1917.[32] The press, commercial organizations, and public service clubs participated actively in the debate.[33] Both sides conducted educational campaigns, sought the support of commercial and manufacturing associations throughout the country, and lined up congressmen. At mass meetings in New Orleans and Memphis, the centers of the opposing forces, partisans debated the technical merits

[29] For a list of drainage companies with which Maxwell worked, see "Minutes of the 1st Meeting of the Louisiana Reclamation Association, Mar. 24, 1912" (GHM, Minutes #1); see also list of land companies and acreages as compiled by Maxwell in GHM, "Special letters and telegrams #1."

[30] A. L. Arpin to various businessmen, nd (GHM, Special letters and telegrams #1); Resolution of the Louisiana Reclamation Club, "Minutes of the 1st Meeting of the Louisiana Reclamation Club, May 14, 1912" (GHM, Minutes #1). Due to flood damage, some of these companies had difficulty in interesting buyers in their reclaimed land. See Charles Kreis to William B. Reilly, Apr. 22, 1912 (GHM, Special letters and telegrams #1).

[31] Minutes of meeting of the Louisiana Reclamation Club, Mar. 24, 1912 (GHM, Minutes #1); also flier put out by the Louisiana Reclamation Club, May 6, 1912 with names of officers, executive committee, and board of governors (GHM, Minutes #1).

[32] See, for example, "Symposium on the Flood's Lesson," *American Review of Reviews* (June 1913), XLVII, 692–704; Maxwell to Newlands, Aug. 6, 1912 (GHM, Special letters and telegrams #1).

[33] The *New Orleans Item* and its owner, James M. Thomson, actively supported the Newlands bill. The Memphis *Commercial Appeal* supported the levee appropriation; its managing editor, C. P. J. Mooney, was vice-president of the Mississippi River Levee Association.

of their respective plans.[34] At their regular meetings, professional engineers, in a more subdued and dignified atmosphere, differed on the worth of the two schemes.[35]

In this public debate the technical superiority of each plan received most attention. According to the Mississippi River Commission, levees of sufficient height and strength to contain the swollen river provided the safest, surest, and least expensive answer to the flood problem. While admitting that levees in the Middle Valley would raise flood heights by eliminating the natural reservoirs in the swamp basins along the river, the Commission argued that a system of levees throughout the Valley would eventually protect all the flood plain.[36] On the other hand, according to partisans of the Newlands bill, in spite of the levees flood damage had continued at a high level; this, they argued, demonstrated the necessity of other measures. Levees should be replaced by reservoirs and forest cover on the upper watersheds, methods which the Roosevelt administration and Senator Newlands had long emphasized.[37]

To the Middle Valley the Newlands bill seemed far too grandiose and remote. Despite the arguments of their critics, levees had helped to reduce flood damage. Even though greater in height, succeeding floods had produced fewer crevasses and had damaged less land. Moreover, these breaks had not occurred in levees of the standards prescribed by the Mississippi River Commission.[38] The Commission's plans seemed definite and involved only flood protection, while the Newlands proposals contained many elements, not yet clear, which might easily take precedence over flood protection, the major

[34] See newspaper report, paper not named, June 20, 1912, of debate between Maxwell and Ransdell at mass meeting in New Orleans (GHM, Personal and Biographical #1).

[35] *New Orleans Picayune*, July 9, 1912 (GHM, Personal and Biographical #1).

[36] *St. Louis Republic*, March 5, 1914 (GHM, Clippings, v. 17); *Montgomery* (Ala.) *Advertiser*, Apr. 15, 1913 (GHM, Clippings, v. 16).

[37] See memorandum by Maxwell, "The Problem of the Mississippi River" (GHM A 6).

[38] Frank, *Flood Control*, 146.

concern of the Middle Valley.[39] Senator Ransdell emphasized the need for a specific plan: "In order to succeed . . . the people of the lower Mississippi Valley must get behind some definite plan and push it rather than to scatter their efforts upon merely allied projects such as those contained in the Newlands' bill."[48] Partisans of the Newlands measure argued that the entire levee movement served merely as a cloak for the operations of real estate promoters. On the contrary, the strength of the levee proposal lay in the fact that it was limited, tangible, and to the point. It had worked in the past, and might work better if expanded. It made sense to those who more than once had faced flood waters first hand and on their own property.

Progress of the Newlands Bill in Congress, 1911–1920

After Congress had failed to approve his proposal in the spring of 1908, Newlands did not introduce it again for three years. After the Democratic victory in the 1910 congressional election, however, he renewed his efforts. Since the Democratic platform in 1908 had contained a plank favoring his proposal, the Senator now hoped that party leaders would help him obtain full Democratic support for his measure. In the closing days of the 61st Congress he introduced a new version of his plan, and on April 6, 1911 he reintroduced it in the special session of the 62nd Congress. This new bill, which George Maxwell had carefully prepared, was referred to McGee and Pinchot, among others, for criticism.[41] Since Newlands hoped that the Smithsonian Institution would carry out the hydrological investigations which the bill authorized, he persuaded its

[39] *New Orleans States*, May 18, 1913 (GHM, Clippings, v. 16); statement of Major Kerr at the Progressive Union meeting, Feb. 21, 1912 (GHM, Correspondence, 1911–July 1914).

[40] Memorandum of the meeting of the Mississippi River Levee Association, Sept. 25, 1912 (FGN, Waterways-River Regulation, Correspondence, 1913, 2).

[41] Maxwell to Newlands, Jan. 5, 1911 (FGN, Waterways-River Regulation, Correspondence, 1907–1911, 2); handwritten note by Maxwell; McGee to Maxwell, Feb. 10, 1911; Pinchot to Maxwell, Feb. 15, 1911 (all in GHM, Clippings—Newlands Bill).

director, Charles D. Walcott, to go over the measure carefully. Walcott, in turn, strongly supported the plan, and urged both Cannon and Taft to adopt it as an administration measure.[42]

In February 1912 the bill was referred to a special subcommittee of the Committee on Commerce, composed of Senators Burton, Crawford, Oliver, Martin, and Newlands. The subcommittee held only one meeting. It agreed on the basic principle of administrative coordination, but differed radically on the size of the appropriation. Newlands argued that Congress immediately should adopt continuing appropriations of $50,000,000 annually for ten years. The majority, however, preferred a sum limited to $1,000,000 for examinations and plans, and Burton demanded that the Commission include two Senators and two Representatives. Although the majority agreed to report the bill if Newlands would agree on these two changes, the Senator from Nevada declined. These were the very modifications, however, to which he later was forced to yield in order to obtain Senate consideration for his measure.[43]

Newlands now tried to attach his plan to the Rivers and Harbors bill, and on April 30, in the Committee on Commerce, offered such an amendment. Favorable in general, the Committee objected to details in the proposal, and begged Newlands not to endanger the entire rivers and harbors measure with his plan. Newlands agreed to withdraw the amendment, while the committee, in turn, agreed to instruct the special subcommittee to present an early report on the Newlands bill itself.[44]

To attract more support, Newlands specified the allotments to be granted to each geographical area. This concession to logrolling ran counter to the basic spirit of the conservation movement. Newlands had attempted to avoid logrolling in the Reclamation Act of 1902

[42] Handwritten note by Maxwell in GHM, Clippings—Newlands Bill; Overton Price to Pinchot, Mar. 18, 1911 (GP #1817).

[43] Newlands to William M. Bunker, Feb. 24, 1912; Newlands to Maxwell, Mar. 2, 1912 (both in FGN, Waterways-River Regulation, Correspondence, 1912, 1).

[44] Newlands to Maxwell, Apr. 2, 1912; Newlands to Maxwell, May 1, 1912 (both in FGN, Waterways-River Regulation, Correspondence, 1912, 2).

by granting the Secretary of the Interior power to locate projects where he thought best. This, however, only transferred the political pressure from Congress to the Secretary. At the direction of President Roosevelt, and in order to satisfy all geographical areas, Secretary of the Interior Hitchcock spaced irrigation projects throughout every Western state.[45] Hoping to avoid these consequences for his waterways bill, Newlands had first proposed that long-term funds be granted to each of the departments or bureaus responsible for water development. But, in 1912 he substituted a geographical distribution of funds in place of allotments to the departments, in order to obtain more congressional votes.[46] His account of this concession revealed the inner workings of the logrolling process:

I also sought to bring together the Senators on the Ohio, the Mississippi and the Missouri Rivers, by this amendment providing for a river regulation board and a continuing appropriation of fifty millions annually from the time of the completion of the Panama Canal, and an intermediate appropriation of five million dollars annually, of which, in addition to the appropriations carried by the river and harbor bill, the Mississippi River from St. Louis down should receive two millions, the Missouri River one million, the Ohio River one million, and the Sacramento and San Joaquin Rivers one million, my idea being to unite the Senators on the Mississippi, the Ohio, the Missouri, and the Pacific Coast in favor of the amendment. But the Committee compromised with the various Senators in other ways, and I found it difficult to unite them in the big legislation.[47]

The Senator hoped, however, to salvage something from this concession. According to his plan, the independent commission would have freedom to choose the specific projects it would undertake

[45] 60th Congress, 1st Session, *Congressional Record*, 395; Roosevelt to Hitchcock, July 2, 1902 (TR). The first twenty irrigation projects authorized from 1903 to 1905 were located in all seventeen Western states except Oklahoma.

[46] Newlands to Maxwell, July 19, 1912 (FGN, Waterways-River Regulation, Correspondence, 1912, 3).

[47] Newlands to Maxwell, May 1, 1912 (FGN, Waterways-River Regulation, Correspondence, 1912, 2).

within the broad limits set by the geographical distribution of funds.

In January 1913, during the third session of the 62nd Congress, Newlands received new support when Secretary of War Stimson and Secretary of the Interior Fisher reported favorably on the measure to the Commerce Committee.[48] Overruling the recommendations of the Corps of Engineers, Stimson forcefully endorsed the plan. He stressed particularly that joint power-navigation works would yield water power proceeds with which the federal government could finance extensive river development projects.[49] This new backing enabled Newlands to persuade the Commerce Committee to report his measure to the Senate, without recommendation, as an amendment to the Rivers and Harbors bill. In this form it passed the Senate. The amendment authorized $500,000 to establish a River Regulation Commission, composed of the Secretaries of War, Interior, Agriculture, and Commerce and Labor, two Senators and two Representatives. This Commission would undertake investigations, draw up plans, and organize a coordinated Water Development Service. To secure favorable action, Newlands had been forced to agree to the changes which the Commerce Committee had demanded a year earlier—the inclusion of Senators and Representatives on the Commission, and the limitation of appropriations to planning. But he felt confident that funds for actual development would follow once the Commission had begun its work.[50] The conference committee upset Newlands' plans by throwing out the entire amendment, and leaving him only the slender satisfaction of

[48] Newlands to Maxwell, telegram, Feb. 14, 1913 (FGN, Waterways-River Regulation, Correspondence, 1913, 1). In his annual report for 1911, Fisher had recommended that proceeds from water power be applied to general waterway development; see Department of the Interior, *Annual Report of the Secretary, 1911*, 1, 15.

[49] Stimson to William C. Adamson, Chairman of the House Committee on Foreign and Interstate Commerce, July 30, 1912 (RG #77, #49678/43); Stimson to Newlands, Jan. 27, 1913 (FGN, Waterways-River Regulation, Correspondence, 1912, 1).

[50] Newlands to Maxwell, telegram, Feb. 25, 1912 (FGN, Waterways-River Regulation, Correspondence, 1913, 1); 62nd Congress, 3rd Session, *House Rept. 1607*, 1, 12.

a Senate report favorable to the measure with which he could further publicize his views.[51]

The Democratic presidential victory in 1912 provided another opportunity for Newlands to strengthen his measure by presenting it as part of the new administration's program. Newlands and Representative Broussard of Louisiana had secured endorsement of the plan in the Democratic national platform in 1912.[52] That September the Senator had persuaded presidential candidate Woodrow Wilson to telegraph the National Irrigation Congress, approving the "policy of supplementing bank and levee protection by storage of flood waters above for irrigation and water power turning floods from a menace into a blessing and at the same time abundantly feeding navigable waters."[53] Newlands' major opponents, however, were also Democrats, and the President soon faced a difference of opinion within his own party which he had to reconcile before he could take up water development as an administration measure.

Joseph E. Ransdell, Democratic congressman from Louisiana, led the drive for a federally financed levee program in the Mississippi Valley. As organizer of the resurrected Rivers and Harbors Congress in 1906, Ransdell stimulated the waterways boom primarily to force the House Rivers and Harbors Committee and the Corps of Engineers to grant his flood control demands. He threatened to transfer the work of the Corps to a civilian agency if it did not change its attitude; however, the Mississippi floods of 1912 and 1913 were far more effective in persuading Congress to undertake such a program. In 1912 Ransdell and Representative Benjamin C. Humphreys of Mississippi presented to the House the program of the Mississippi Valley Levee Association, which called

[51] Newlands to Senator Coe I. Crawford, Feb. 27, 1913 (FGN, Waterways-River Regulation, Correspondence, 1913, 1); 62nd Congress, 3rd Session, *Sen. Rept. 1339*.

[52] Newlands to Maxwell, Mar. 4, 1912 (FGN, Waterways-River Regulation, Correspondence, 1912, 1).

[53] Newlands to Wilson, telegram, Sept. 28, 1912; Wilson to Newlands, telegram, Sept. 29, 1912; Newlands to Wilson, telegram, Sept. 30, 1912 (all in FGN, Irrigation—National Congress, 1909–1915, 3).

for $60,000,000 in federal aid. This measure immediately came into competition with the Newlands bill, and for a number of years the two groups thwarted each other's plans.

To Senator Newlands a program limited to levee construction typified the single-purpose programs which should be coordinated into a multiple-purpose plan. He found no fault with levees; he merely felt that Congress should tackle the flood problem from every feasible angle. The Ransdell-Humphreys group, on the other hand, feared that their levee program would suffer from neglect if it were included in a comprehensive river development plan. An executive commission might not respond to the pressures of the Mississippi Valley as readily as would a congressional committee. Finally, while the Newlands measure might involve preliminary investigations for a number of years, the Valley demanded immediate relief.

President Wilson tried to reconcile the views of his two Democratic leaders. In November 1913 they agreed to submit their proposals to an interdepartmental committee which, they hoped, would formulate a single proposal, agreeable to both sides, to be presented to Congress as an administration measure.[54] Preparing a preliminary report early in January 1914, and a more complete statement in April, the committee's majority preferred the Newlands approach. It looked with special favor on the principle of administrative coordination and the appropriation of lump sums to an executive agency which could spend them on a continuing-contract basis.[55] This proposal hardly provided a common ground acceptable to Senator Ransdell. But, when the Rivers and Harbors Committee of the House turned down his flood control measure, the Senator from Louisiana felt compelled to work out some agreement directly

[54] Newlands and Ransdell to Secretaries of War, Interior, Agriculture, and Commerce, Dec. 9, 1913 (RG #77, #83879/81).

[55] "Report of the Inter-Departmental Committee on the Newlands Bill," Jan. 10, 1914 (RG #77, #83879/142); Memorandum of the Committee, Jan. 16, 1914 (RG #77, #83879/197); "Supplemental Report of the Inter-Departmental Committee on the Newlands Bill," March 30, 1914 (RG #77, #83879/200).

with the Senator from Nevada to save his levee program. He offered to join with Newlands if the measure included a specific appropriation for levee construction on the Mississippi and left the Corps of Engineers free to administer the program.[56] Newlands accepted the arrangement, and merely incorporated the Ransdell-Humphreys measure into his plan.[57] At the same time, the opposing political groups in the Mississippi Valley joined to support both the Newlands bill and the levee appropriation. The stage was set for the administration to act.

The Corps of Engineers, it alone not a party to the compromise, almost disrupted this strategy. Secretary of War Lindley M. Garrison refused to introduce the new version of the bill as an administration measure, and the cabinet on May 29, 1914 failed again to press the matter.[58] A vigorous attack on the Rivers and Harbors bill, however, prompted the cabinet to change its decision.[59] In the House, James A. Frear of Wisconsin led the onslaught on the logrolling which had produced no funds for the Mississippi River section of his state. On May 5 he introduced a resolution calling for an Interstate Commerce Commission investigation of the entire River and Harbor program. In the Senate, Burton, aided by Kenyon, Norris, and Ashurst, continued the assault. Burton, still trying to quell the enthusiasm for raiding the public coffers, received new support from Western Senators, led by William E. Borah of Idaho, who maintained that Southern Democrats had written the measure and had ignored the just claims of Western irrigators for more funds.[60]

In the face of the charge that the Democrats had brought forth a monstrous "pork barrel" in the Rivers and Harbors bill, the ad-

[56] Maxwell to M. F. Hudson, Feb. 12, 1914 (FGN, Waterways-River Regulation, Correspondence, 1914, 1).

[57] S 2730, introduced Jan. 31, 1914.

[58] Newlands to Garrison, May 22, 1914 (RG #77, #83879/156); Garrison to Secretary of Interior Lane, May 29, 1914 (RG #77, #83879/163).

[59] This attack on the rivers and harbors bill can be followed in Pross, "Rivers and Harbors," 182 ff.

[60] *New York Herald*, May 5, 1914 (GHM, Clippings, v. 19); *New York City Press*, June 10, 1914 (GHM, A 7).

ministration sought to restore its respectability by backing the Newlands measure.[61] On June 22 the Senator introduced the same proposal for an investigating commission which the Senate had approved the year before, and attached it to the Rivers and Harbors bill. Newlands wrote Maxwell: "It has, as stated, the full approval of the President and the Cabinet, and the Democratic leaders have agreed to let it go through."[62] Confident that his amendment would pass, Newlands busily worked out the organization and personnel of the future commission.[63] But he planned too soon. The battle over the Rivers and Harbors measure continued. The administration grew weary of the debate, dropped the Newlands amendment, and accepted a measure pared down to a lump sum of $20,000,000, which the Corps of Engineers could apportion as it saw fit.[64]

In the 64th Congress the Wilson administration returned to the Newlands plan. In the spring of 1916 Newlands, Ransdell, and Broussard persuaded Secretaries Lane, Houston, and Redfield, of the Interior, Agriculture, and Commerce departments respectively, to recommend that Wilson send a special message to Congress.[65] When Secretary of War Baker came into the cabinet a short time later, he enthusiastically endorsed the Newlands bill. Later he explained, "The uses of water were so closely related one with the other that it was highly desirable that there should be some coordinating influence in their consideration and study from the administrative and executive side, and so I found myself in perfect agreement with the plan that Senator Newlands now has."[66]

[61] Maxwell to Newlands, June 6, 1914; Maxwell to Newlands, June 10, 1914 (both in FGN, Waterways-River Regulation, Correspondence, 1914, 4).

[62] Newlands' secretary to Maxwell, June 24, 1914 (FGN, Waterways-River Regulation, Correspondence, 1914, 4); Garrison to Senator F. M. Simmons, June 23, 1914 (RG #77, #83879/177).

[63] Newlands to Maxwell, June 29, 1914; Newlands to Maxwell, Aug. 10, 1914; M. F. Hudson to Maxwell, Sept. 16, 1914 (all in FGN, Waterways-River Regulation, Correspondence, 1914, 4).

[64] 63rd Congress, 1st Session, *Congressional Record*, 15873, 15931.

[65] Newlands to Maxwell, Mar. 1, 1916 (FGN, Waterways and Irrigation—Correspondence, 1913–1917, 4); the three Secretaries to the President, Feb. 26, 1916 (RG #77, #125806/1).

[66] *Washington Star*, Dec. 23, 1916 (GP #1665).

The cooperation of Ransdell and Humphreys proved to be even more important to Newlands than did Wilson's halfhearted aid. Humphreys finally persuaded the House to appoint a Flood Control Committee, with himself as chairman, which reported his bill favorably. Passing the House, it came up in the Senate in the summer of 1916. Here Ransdell and Humphreys confronted the prestige and influence of Senator Newlands. They agreed with the Senator from Nevada to introduce the flood control bill as a part of the broader river development measure.[67] In December 1916, however, they shifted their tactics and persuaded the Senate Committee on Commerce to report the original Humphreys measure alone. Indignant at this maneuver, Newlands blocked the bill on the Senate floor.[68] The Mississippi Valley Senators then came to terms and agreed to support the Newlands plan as an amendment to the 1917 Rivers and Harbors bill, if Newlands in turn would back the flood control proposition. The bargain was fulfilled; the flood control bill passed on March 1, 1917, and on August 8, the Rivers and Harbors Act, with the waterway commission plan as Section 18, became law.

Appointments to the waterway commission gave rise to a controversy as bitter as that over its authorization. Newlands and Maxwell argued for a personnel of cabinet members, who could delegate responsibility to subordinates and base their recommendations on technical data.[69] Senator Ransdell and his friends, however, preferred a non-cabinet group which would respond more readily to Congress. They feared that cabinet members might coordinate the work of the Army Engineers with that of other agencies and en-

[67] Ransdell to Newlands, Nov. 14, 1916; V. S. McClatchy to Lane, Dec. 12, 1916; Newlands memorandum, Dec. 16, 1916 (all in FGN, Waterways—Flood Control, 1916, 3).

[68] McClatchy to Lane, Dec. 12, 1916; Lane to Newlands, Dec. 14, 1916; Newlands memorandum, Dec. 16, 1916 (all in FGN, Waterways—Flood Control, 1916, 3).

[69] Newlands to President Wilson, Aug. 20, 1917 (FGN, Waterways—Flood Control, 1917, 1).

danger the flood program.[70] Major-General William M. Black, Chief of the Corps, supported this view. Congress, he argued, did not wish a cabinet commission, which would be unwieldy and unworkable.[71] Preoccupied with the War, neither President Wilson nor Secretary of War Baker could take time to resolve these conflicting interests; the commission was never appointed.

This turn of affairs greatly embarrassed the advocates of multiple-purpose development. The death of Senator Newlands in 1919 deprived the movement of its most forceful and strategic figure. Determined to destroy what they had been forced to approve in 1917, the foes of Newlands sought to repeal Section 18 of the 1917 Rivers and Harbors Act. Maxwell worked frantically to prevent this. He first proposed a substitute amendment which would name specifically the Commission's members: the Secretaries of War, Interior, and Agriculture, the Directors of the Geological Survey and the Bureau of Reclamation, the Chief of the Forest Service, and the Chief of the Corps of Engineers. The Senate agreed to this proposal and then refused to appropriate funds for the Commission's work. Maxwell did not despair; he hoped that the group as thus constituted could make considerable progress without special appropriations, but with only departmental funds. Moreover, he felt that once the Commission had begun work it could easily secure appropriations for itself.[72]

Maxwell, however, worked in vain. In the Water Power Act of 1920 Congress not only repealed Section 18 without qualification, but also sealed the fate of multiple-purpose river development by enacting a law which violated those very principles.[73] Admittedly, the Water Power Act established public supervision of hydroelectric development on both the public lands and the navigable streams,

[70] Newlands to Wilson, Sept. 24, 1917 (FGN, Waterways and Irrigation, Correspondence, 1913–1917, 5).

[71] Major-General William M. Black, "Memorandum for the Secretary of War," July 2, 1918 (GHM, A 7).

[72] Maxwell to W. S. Wright, Jan. 15, 1920; Maxwell to Julian Kennedy, Jan. 20, 1920 (both in GHM, A 7).

[73] 66th Congress, 2nd Session, *House Rept. 910*, 13–14.

but it failed to provide for multiple-purpose development as the goal of public action. It omitted flood control and irrigation and spoke of navigation only in the ancient language of the Corps—that hydroelectric installations should not interfere with it. The law failed to link power revenues with construction of multiple-purpose dams. It provided for only meager financial returns to the federal government from hydroelectric installations, and, in fact, reduced the revenue from the levels enjoyed by the Forest Service alone under its permit system. Despite these omissions, conservationists looked upon the Water Power Act as a fair compromise with the private power corporations. On the contrary, that Act, representing, finally, an adverse decision on the multiple-purpose idea, constituted a defeat and marked the end of a conservation era. Weary from a decade of debate, the Roosevelt leaders accepted the symbolic action of public regulation instead of the substance of the multiple-purpose principle itself.

Chapter XII

The West Against Itself

A comprehensive water development program failed because many feared that executive planning would exclude them from influence in resource policy. Scientific management of the Western public domain provoked similar opposition. Although many Westerners sought and obtained federal assistance, they resisted the controls that accompanied it, and complained that they could not influence the decisions of administrative agencies. In the vast geographical area of the West, moreover, a variety of conditions gave rise to disagreements over Western policies. Since aid to one group frequently was detrimental to another, federal resource programs became intimately involved in these intra-Western conflicts, and one group, in turn, transferred its hostility toward another to a federal scapegoat. Many intra-Western disagreements in this way became controversies between the West and Washington. "Divided against itself," the West displayed a fierce independence, yet courted the intervention of forces which it could not always control.

Federal or Private Irrigation?

The federal irrigation program displeased many Westerners who believed that development should proceed under private rather than federal auspices. Private water users feared that federal agencies would interfere with water rights established under state law, and private promoters were concerned lest the federal government restrict private undertakings. These objections received little support in face of the great need for capital. Yet, after the Reclamation Act of 1902 became law, the same misgivings persisted.

During the first few years of the federal reclamation program promoters of both private and public projects frequently desired to use the same reservoir location or the same lands for irrigation.[1] The Act of 1902 gave the Bureau of Reclamation the upper hand over private parties by authorizing the Secretary of the Interior to withdraw prospective reservoir sites and irrigable lands from entry. At the insistence of Frederick H. Newell, Chief of the Bureau, the Secretary withdrew lands which the Bureau could not develop immediately, but might be able to take up in the future when funds became available. After investigating the Central Valley in California and the Colorado River, for example, the Bureau formulated plans for large-scale development in both areas which it could carry out eventually if it could retain control of the strategic lands involved. Private entrepreneurs, desiring to use the same sites, applied for rights-of-way, but were told that their projects would interfere with the Bureau's plans.[2]

Most federal projects brought water to private as well as public lands, thereby setting the stage for controversy between private irrigators and the Bureau. Property owners complained that the Bureau would not pay enough for the lands and canals which it purchased for its own use.[3] They also protested bitterly against the provision of the Reclamation Act of 1902 that no private land owner could receive rights to water from a federal project for more than 160 acres.[4] Under this restriction, the Bureau refused to under-

[1] *Irrigation Age* (Jan. 1909), 24, 73–74.

[2] T. C. Henry, "The Reclamation Service versus the State of Colorado," *Proceedings, Trans-Mississippi Commercial Congress, 1909*, 83–89. In March 1903 a bill was introduced in the House to authorize the president to restore withdrawn reservoir sites. See the *Forester* (Mar. 1903), 9, 151–152.

[3] *Salt Lake Herald*, June 28, 1903 (FN #17, Clippings, v. 4). In the case cited it was reported that land purchased for $40,000 was offered to the Bureau of Reclamation for $300,000; when told that their holdings would be condemned the syndicate owning the land lowered its price to $200,000.

[4] Provision for private land irrigation from federal projects was placed in the Act of 1902 only during the latter parts of its legislative history, almost as an afterthought. See statement of Newell in *Proceedings, National Irrigation Congress, 1905*, 27.

take a project unless it first held written agreements from all who owned more than 160 acres that they would not sell excess land at inflated prices. This 160-acre law provoked little opposition in the Southwest where the water shortage was acute. For example, at the Salt River Valley project in Arizona, involving private land alone, irrigators readily accepted the limitation. In the Pacific Northwest, however, an area of more plentiful rainfall, severe objections arose. Projects in Oregon were held up for many years because property owners would not agree to sell at nonspeculative prices.[5]

Private irrigators found vigorous spokesmen who carried on the attack against the Bureau of Reclamation. Even prior to the passage of the Act of 1902, the *Irrigation Age*, a monthly periodical, had taken up their cause and warned that the West would regret federal aid. It continued the attack throughout the early years of the twentieth century.[6] Criticism of the federal program gathered behind the figure of Elwood Mead, the former Wyoming state engineer, who was concerned especially with state water law problems. In 1898 Mead became Chief of a new Office of Irrigation Investigations in the Office of Experiment Stations in the Department of Agriculture. In response to the demands of those who favored state development, Congress had established this office after curtailing the work of the Geological Survey, the major federal agency in the 1890's to back a federal aid program. Mead and his fellow scientists sought to foster private irrigation by providing technical aid and by encouraging states to develop a more rational water law.[7]

The Office of Irrigation Investigations and the Geological Survey vied to administer the new program established by the Newlands Act of 1902. Knowing of Mead's hostility to the program, Pinchot and Newell persuaded Roosevelt to entrust the Survey with the

[5] *Los Angeles Express*, Oct. 4, 1905 (FN #17, Clippings, v. 5); *Maxwell's Talisman* (Oct. 1905), 5, 3; *Irrigation Age* (Mar. 1910), 25, 198–200.

[6] George H. Maxwell to Pinchot, June 30, 1903 (GP #80); *Irrigation Age* (Feb. 1909), 24, 106–107; (Jan. 1909), 24, 86–87.

[7] Newell to Pinchot, July 2, 1903 (GP #1937); A. C. True to Pinchot, Dec. 10, 1903 (GP #1937); Mead to Pinchot, July 2, 1903 (GP #1937).

task.[8] Although Mead thereafter expressed approval of the federal program in principle, he frequently criticized it in practice in the columns of *Irrigation Age* and before irrigation conventions. Those who criticized the work of the Bureau of Reclamation looked upon the Office of Irrigation Investigations as a rival body far more friendly to private irrigation development.[9]

Newell, the Geological Survey, and the Bureau of Reclamation returned the criticism, especially after 1902 when the *Yearbook of Agriculture* contained articles hostile to federal irrigation and friendly to private development.[10] The Bureau became even more irked when its Western opponents sought to increase their influence by claiming that the Department of Agriculture, through its Irrigation Office, supported their view;[11] Bureau officials came out in a frontal attack on Mead's policies. State engineers, they argued, shifted with the whims of state politics. Western states should settle their water conflicts through court, rather than administrative action.[12] Using the President's Committee on Scientific Work in Government as the springboard for his attack, Newell tried to abolish the Office of Irrigation Investigations. To that Committee he complained that the Office hampered his efforts by creating an opposing political group. He recommended that its functions be transferred to other agencies.[13] But Congress had checked Newell's proposal even before it was made. For reasons still unknown the law under which the President could consolidate scientific work in the federal administration had excluded from its provisions agencies in the Department of Agriculture.[14] The Committee on Scientific Work proposed special legislation to abolish the Office of Irrigation Investigations, but Congress took no action.

[8] Roosevelt to Hitchcock, June 17, 1902 (TR); *Washington* (D.C.) *Times*, July 2, 1902 (FN #16, Clippings, v. 3).

[9] *Ibid*; *Portland Telegram*, Nov. 7, 1902 (FN #16, Clippings, v. 3).

[10] *Washington* (D.C.) *Times*, July 2, 1902 (FN #16, Clippings, v. 3).

[11] Newell to Pinchot, July 2, 1903 (GP #1937).

[12] *Denver Republican*, Sept. 21, 1903 (FN #17, Clippings, v. 3); Charles D. Walcott to Newlands, Dec. 11, 1902 (FGN, Letters).

[13] Newell to Pinchot, July 2, 1903 (GP #1937).

[14] Willis L. Moore to Pinchot, July 16, 1903 (GP #1936).

The National Irrigation Congress, organized to support the federal program, became a battleground in the struggle between private and public promoters. Each year at the convention of the Congress, representatives of private irrigators introduced resolutions censuring the Bureau of Reclamation. Federal officials, in turn, worked closely with their Western friends to defeat these moves. In 1905 the enemies of the Bureau did succeed in persuading the Congress to disavow all connection with George Maxwell, who, though not officially connected with the federal program, had become closely associated with it in the public mind. Their efforts to censure the Bureau itself, however, failed.[15] Shortly after the 1905 meeting, the dissidents, under the leadership of Grant L. Shumway of Scottsbluff, Nebraska, organized their own group—the American Irrigation Federation.[16] The editor of *Irrigation Age* became secretary of the Federation and one of the vice-presidents was Clarence T. Johnston, Wyoming state engineer who for several years had worked in the Office of Irrigation Investigations as assistant to Elwood Mead. The American Irrigation Federation undertook a widespread campaign against federal irrigation, but it made little headway and by 1911 had disappeared.

The Federal Government as Creditor

As the Bureau of Reclamation completed its project planning and began to construct reservoirs and ditches, its conflicts with private promoters declined. At the same time, even more prolonged difficulties arose with the settlers who hoped to receive water from the federal works. Homesteaders quickly took up vacant lands under federal projects, even before construction had begun, and exerted constant pressure on the Bureau to complete the work rapidly. But construction proceeded slowly. In response to Western demands, the administration had started work in every major state of the region, thereby spreading out the funds and slowing up progress on any

[15] *Los Angeles Times*, Sept. 3, 1905 (FN #17, Clippings, v. 5); C. B. Boothe to Pinchot, Sept. 4, 1905 (GP #2138).

[16] *Irrigation Age* (Mar. 1909), 24, 133–134; (May 1909), 24, 206–207.

one project.[17] To speed the work, Newell originated the scrip labor device which the Taft administration declared illegal. The West then recommended that Congress provide a special bond issue to complete the projects. At first Ballinger denied that the Bureau needed more funds, but he and Taft finally agreed to a $20,000,000 issue which Congress approved in 1910.[18]

A thornier problem arose when the settlers declined to follow the Bureau's plans for repaying construction funds.[19] Under the Reclamation Act the settlers agreed to pay for the projects by returning the investment, though without interest, to the Reclamation Fund in ten annual installments. As the time of initial repayments neared, settlers on almost every project requested more lenient terms, protesting that the Bureau calculated the returnable costs at far too high a level. The Act of 1902, they argued, had provided for repayment of the "estimated project cost," while the Bureau demanded repayment of the final cost, a much higher figure. The Bureau replied that the water users themselves had requested changes, thereby increasing the cost. The water users vowed that they would pay only the original "estimated cost," and demanded that the ten-year repayment period be extended to twenty. Senator Borah of Idaho even introduced a measure to relieve water users of all financial obligations under the Reclamation Act.[20]

These demands came into conflict with the interests of other Western areas which hoped to obtain authorization for additional federal projects. Under the theory of the Reclamation Act, public land proceeds would constitute a revolving fund. Repayments into the Reclamation Fund would provide capital for new projects. Newlands, Pinchot, and Newell all argued that the Bureau should

[17] Roosevelt to Hitchcock, July 2, 1902 (TR).

[18] *Chicago Drovers Journal*, Aug. 12, 1909 (FN #18, Clippings, v. 7); *Irrigation Age* (Nov. 1909), 25, 11; (Dec. 1909), 25, 79; (May 1910), 25, 338.

[19] Repayment problems are fully covered in Peggy Heim, "Financing the Federal Reclamation Program, 1902 to 1919: The Development of Repayment Policy," doctoral dissertation, 1953, Columbia University.

[20] *Phoenix Gazette*, May 22, 1913 (GHM, Clippings, v. 16); *Salt Lake Telegraph*, Sept. 8, 1909 (FN #18, Clippings, v. 7).

adhere to the original repayment schedule, both as a just financial arrangement, and in order to bring irrigation to new areas. At a Salt Lake City Irrigation Congress in 1912, the Senator warned:

It must be recollected that the existing projects are the favored ones entered upon in preference to other projects equally demanding governmental consideration and that the extension of the time of payment beyond the ten years originally contemplated will unjustly delay the inauguration of new enterprises.[21]

Most of the areas not yet irrigated, however, soon despaired of obtaining capital from the Reclamation Fund, and supported the Newlands waterway bill which provided for federal bond issues for new projects.[22]

The drive to change the repayment provisions of the Reclamation Act was spearheaded by a National Water Users Association, organized in 1911. Frederick Newell of the Bureau of Reclamation became the special target of its criticism.[23] Attacks on Newell and his view that repayment should proceed on schedule arose first in the summer of 1909. That year a congressional committee investigating the reclamation program served as a sounding board for the Bureau's opponents.[24] The Committee's majority, however, returned a report friendly to Newell and effectively sidetracked the criticism. President Taft also refused to yield to the irrigators' demands and argued that repayment should proceed as planned. He yielded to the water users only to the extent of backing the $20,000,000 bond issue to complete construction. Taft might have gone further had political considerations not entered the picture. Secretary of the Interior Walter Fisher, irked with Newell's actions, wanted to fire him. Pinchot, however, implored Taft to prevent it, and Taft, desperately trying to mend the breach in the Republican Party, blocked the dismissal.

[21] Speech of Newlands at Salt Lake City Irrigation Congress, Sept. 30, 1912, MSS in FGN, Irrigation—National Congress, 1910–1912, Documents, 3.

[22] The Imperial Valley, for example, became an ardent backer of the Newlands plan. See *Los Angeles Examiner*, Apr. 2, 1914 (GHM, Clippings, v. 17).

[23] George J. Scharschug to the *New York Times*, in *New York Times*, Feb. 14, 1915, II, 8:7. [24] *Irrigation Age* (Sept. 1909), 24, 402; (Oct. 1909), 24, 447.

From the Woodrow Wilson administration the water users obtained a far more sympathetic ear. A Democratic president felt no obligations toward irritating Bull Moosers. Representatives of the Water Users Association immediately placed their case before the new Secretary of the Interior, Franklin K. Lane, during a long meeting in Washington in 1913.[25] Lane, in turn, reorganized the Bureau of Reclamation by placing a commission of three men in charge of its operations, retaining Newell nominally as director, but reducing him to the position of chief engineer.[26] Lane publicly praised the Roosevelt administration's construction work in reclamation, but criticized its attack upon the "human problems" involved. "They did not realize," he wrote, "that, primarily, these lands were being reclaimed for human occupation. They were interested chiefly in making wonderful dams and reservoirs—not in making the people industrious and contented."[27] Lane endorsed and supported the major demands of the water users. The administration-backed Reclamation Extension Act of 1914 extended reclamation payments from ten to twenty years, and transferred the power to allocate the Reclamation Fund to specific projects from the Secretary to Congress itself. Roosevelt and Pinchot bitterly criticized these changes in policy as a product of "Democratic politics" and the influence of "special interests." They became even more indignant when Lane in 1915 dismissed Newell from all connection with the Bureau of Reclamation.[28]

The Politics of Planning

How should conflicts among Western resource users be settled? This problem illustrated concisely the ambivalent Western attitude toward the federal conservation program. To Roosevelt administrators its solution lay at the heart of the conservation idea. Com-

[25] For items relating to this meeting see National Archives, Bureau of Reclamation, Record Group #115, General File 131–A2.

[26] Harry A. Slattery to Overton Price, May 29, 1913 (GP #1828).

[27] *New York Journal of Commerce*, Aug. 13, 1913 (GHM, Clippings, v. 16).

[28] Edward Breck to Slattery, July 20, 1915 (GP #1829); Slattery to Breck, July 15, 1915 (GP #1829).

peting claims to resources should be resolved, they argued, by a scientific calculation of material benefits rather than through political struggle. How else could one guarantee maximum efficiency? Westerners did not agree. To them the "scientific expert" became simply a "bureaucrat" whose decisions a local group could rarely affect. Open economic competition, court action, or political pressure provided far more effective opportunities for resource users to fight their own battles. The cry of "dictator" rarely arose until a federal bureau decided an important case in favor of one group and against another. But an accumulation of discontent in specific cases grew into a widespread regional attack against the "undemocratic methods" of the federal experts.

Water controversies provoked special bitterness toward the Bureau of Reclamation. The Bureau thought in terms of maximum development of all Western rivers, while the people of each local area considered their own welfare first. Plans formulated by the Bureau frequently called for reservoirs in the tributaries to store excess runoff until it could be used during the dry season on lands further down the river. Upper river residents, however, resented the use of "their" water elsewhere, frequently in another state. For example, the Bureau planned a series of reservoirs in the upper Rio Grande River in Colorado to provide water for lands in Texas and New Mexico. Residents of southern Colorado, however, hoped to use the same sites to store water for their own use. The Bureau of Reclamation turned down their applications for rights-of-way on the ground that their plans would interfere with the larger Rio Grande project. The conflict actually involved disagreements between upper and lower Rio Grande water users, but the Bureau was blamed, and Colorado groups demanded that it leave them alone to develop their own state.[29]

The Colorado River involved an even more prolonged conflict. The Bureau's long-range plans for the Colorado called for storage reservoirs in the upper river to irrigate several million acres along

[29] *Proceedings, National Irrigation Congress, 1904*, 107–110; *Proceedings, National Conservation Congress, 1913*, 81–91; *Proceedings, Trans-Mississippi Commercial Congress, 1909*, 84.

its lower reaches in California and Arizona. To further this proposal, the Secretary of the Interior withdrew from entry two reservoir sites, one at Brown's Park on the upper Green River and one in Gore Canyon, on the main river near Kremmling, Colorado. These sites, Colorado residents argued, were needed for that state's development. The Denver, Northwestern and Pacific Railroad, constructing a line from Denver to Salt Lake City, hoped to locate its route in Gore Canyon through the reservoir site. Pressures on Washington revealed clearly the conflicting interests. Colorado congressmen supported the railroad, while the Los Angeles Chamber of Commerce argued forcefully for the reservoirs. The railroad won.[30]

Interstate water conflicts continually beset the West. The Bear River, flowing through Wyoming, Idaho, and Utah, was subject to the laws of three different states. The Little Snake River crossed the Colorado-Wyoming boundary four times.[31] President Roosevelt explained the predicament of the federal government in such cases: "In the Reclamation Service I am obliged to move carefully because of the interstate jealousies over the distribution of waters, and I do not like to take action which may seem to be against the interest of one state or to sacrifice such interest to the interest of another state unless the action is clearly demanded."[32] While the West tried to defend its position through the federal courts and by taking from the Secretary of the Interior his power to approve projects, federal officials sought to solve the problem by developing a national water law to supersede state statutes. In 1907 the National Irrigation Congress advocated such a measure, and Newell continued to support it.[33] In 1915 he persuaded the American Society of Civil Engineers to appoint a special committee to study the question. It found one hundred important interstate water conflicts in the West,

[30] *Forestry and Irrigation* (June 1904), 10, 112–116, 274–278; *Maxwell's Talisman* (Oct. 1905), 5, 8; Roosevelt to Hitchcock, May 26, 1905 (TR); Roosevelt to Philip Stewart, Oct. 9, 1905; May 17, 1906 (TR).

[31] *Conservation* (Jan. 1909), 15, 10.

[32] Roosevelt to William E. Curtis, May 19, 1906 (TR).

[33] *Conservation* (Jan. 1909), 15, 9.

which, it argued, could be solved by a national water law.[34] However, this approach could make little headway in the face of the particularist views of many Western communities who feared that a federal law would further limit their ability to control their most vital necessity of life.

A rational land management program similarly became involved in internal Western rivalries. In this case, the administration's plan to classify and reserve public lands for a particular use aroused intense criticism. The West preferred a less rigid plan that would enable Westerners to choose how land should be used. J. D. Whelpley, a newspaperman familiar with Western conditions, best expressed this point of view:

> . . . all idea of a strict classification of the public lands should be abandoned. . . . The desert of yesterday is the productive irrigated farm of today. Live-stock range or grazing land is constantly being brought into the cultivated area. The classification of the public lands has been urged by eminent authorities, but those who are familiar . . . with the remarkable results of the struggle for homes in the West during the past few years are firm in the belief that not for a generation at least can any man dare say this or that piece of land is uninhabitable or impossible of productiveness.[35]

The Public Lands Commission, which Whelpley himself first suggested to Pinchot to investigate land frauds, laid the plans for systematic development through land classification. But Western farmers would not even accept the Commission's proposal that the classification of farm and range land be adjusted periodically. To the West classification approximated withdrawal. In both cases federal administrators, rather than Westerners themselves, would decide resource use.

The thousands of settlers pouring in to the upper plains and intermountain states in the early twentieth century provided major opposition to planned land development. Attracted by the new "dry

[34] "Report of the National Water Law Committee of the American Society of Civil Engineers," in FN #7–8.

[35] J. D. Whelpley, "The Nation as a Land-Owner," *Harper's Weekly* (Nov. 30, 1901), 45, 1204.

farming" publicity, these homesteaders became convinced that the arid lands could become permanently productive farms.[36] Pinchot and his friends hardly mentioned this movement in their speeches, reports, or private letters. At the same time, they advanced the argument, odious to the settlers, that the remaining public domain could support grazing alone, and therefore should be leased. Many Western settlers, who had supported the administration's irrigation and forest policies prior to 1906, wrote in protest as Roosevelt, Pinchot, and Garfield continued to work with the cattlemen for a leasing law.[37] Frank Mondell, president of the Dry Farming Congress, which spearheaded the drive to popularize farming in the arid country, became the bitterest critic of the administration's public lands policy. Mondell countered the leasing measure with a 320-acre homestead law which would permit land entries more adequate for dry farming. In 1909, having rejected the leasing measure, Congress approved the Mondell bill.

The Federal Landlord

Permanent federal ownership of Western lands set the stage for controversy between landlord and tenant. How much, for example, should be charged for the sale or lease of forest, range, or mineral products? In the past, conservationists argued, the federal government had disposed of its resources far too lavishly. It should now require rentals and fees equal to those charged by private owners. The Roosevelt administration, adopting this point of view, instituted grazing and water power fees and raised prices for timber and minerals. Lumbermen, coal consumers, cattle and sheepmen, and hydroelectric power promoters arose in arms against the new policy. Increased charges, they complained, raised their costs until they could not compete effectively with other Western corporations.[38] Federal fees, the administration replied, should not inhibit legitimate development, but in order to prevent speculation, should be raised

[36] Mary W. M. Hargreaves, "Dry Farming Alias Scientific Farming," *Agricultural History* (Jan. 1948), 22, 39–55.

[37] Paris Gibson to Harry A. Slattery, Apr. 8, 1912 (GP #1817).

[38] Frank Mondell to Walter L. Fisher, Aug. 21, 1912 (GP #1672).

over past rates.[39] Disagreements between landlord and land user plagued federal officials from the date that they increased user fees. Dissatisfaction over the new financial policies created much of the Western opposition to the conservation movement.

When the United States Forest Service began to sell timber at competitive bidding rather than at appraised prices,[40] thereby increasing the cost of stumpage, the few lumbermen who used the reserves could arouse little interest in their protests.[41] The charge for grazing, however, involved a great number of Westerners who could make themselves heard. Prior to 1905 stockmen had grazed on the public lands without charge. While lumbermen faced only a rise in prices, grazers were confronted with fees for the use of forage for the first time. When the Forest Service announced the new policy in the summer of 1905, the industry reacted quickly and violently. A Wyoming editor expressed the stockmen's attitude:

> We have, struggling with frontier conditions, built up the sheep and cattle interests of this section under many discouragements and have barely been able to win a modest success. . . . The free and unrestricted use of this mountain range has always been enjoyed by us. We have bought our property and built up our ranches on the strength of its continued use without tax or charge. Aided by this public range, under present conditions, we have been hardly able to hold our own, and the imposition of any grazing tax means to us disaster and ruin.[42]

To the administration it was "perfectly obvious that the man who pastured his stock should pay something for the preservation of that pasture."[43] But to the stockmen it was not so obvious.

The Forest Service fortified its position when George Woodruff, its law officer, obtained from Attorney-General Moody the ruling that it had power to charge a fee.[44] To reduce opposition to the plan,

[39] Fisher to Mondell, Aug. 17, 1912 (GP #1672).

[40] *Forestry and Irrigation* (June 1907), 13, 278.

[41] R. D. Meyer to Senator Francis E. Warren, June 9, 1906 (GP #1715).

[42] *Centennial* (Wyo.) *Post,* Dec. 16, 1905, as quoted in Robbins, *Our Landed Heritage,* 348.

[43] Roosevelt to Secretary of Agriculture James Wilson, Dec. 21, 1905 (TR).

[44] Pinchot, *Breaking New Ground,* 271–272.

Pinchot required only half-rates of small ranchers for the first year. But stockmen did not accept the new policy without vigorous protest. Some talked of "taxation without representation," and one journal proposed that the Western states secede from the Union. Senator Thomas M. Patterson, a Democrat of Colorado, led the attack. In December 1905 he complained to President Roosevelt by letter and later repeated his arguments in person.[45] The Denver Public Lands Convention in June 1907 served as an opportunity for the entire range country to attack the new policy. In fact, the Colorado legislature called this meeting specifically to protest the grazing fees as well as to object to the plan to enact a general leasing measure for the entire public domain.[46] When these efforts did not budge the Forest Service, stockmen instituted test cases in the courts to challenge its authority to regulate grazing on the national forests in any form. In 1911, however, the Supreme Court decided each contest in favor of the federal government.[47]

Accepting the Supreme Court decision, stockmen now tried to keep the fees as low as possible. The Forest Service charged only one-third the amount of corresponding rentals on privately owned grazing land.[48] Its officials hoped to raise the fees, because it seemed only just to bring public charges in line with private, and because increasing revenue from the forests might persuade Congress to grant greater appropriations for range protection. They also fully appreciated the political danger of raising fees. Action within the Forest Service to increase charges began as early as 1911, but the grazing division successfully prevented the move until 1916.[49] In that year the Secretary of Agriculture announced an increase of 100 per cent, which, he argued, would bring public rentals into line with private. Stockmen protested vigorously, but to no avail.[50]

[45] Roosevelt to Thomas M. Patterson, Dec. 21, 1905 (TR).

[46] Lute Pease, "The Way of the Land Transgressor," in *Pacific Monthly* (Aug. 1907), 18, 145–164 (FGN, Scrapbook #23).

[47] Wells, "Personal History" (GP #1671).

[48] Overton Price to Pinchot, Apr. 15, 1911 (GP #1817).

[49] *Ibid.*; Price to Pinchot, Apr. 22, 1911 (GP #1817).

[50] Memorandum, Harry A. Slattery to Pinchot, Nov. 18, 1916 (GP #1834).

Western congressmen protested just as vigorously when the administration raised coal land prices. Representative Mondell of Wyoming led the opposition, which arose primarily in the major coal-producing states of Wyoming and Colorado.[51] By raising the price of coal to mining companies, Mondell argued, the new policy retarded land sales, raised retail prices, and promoted monopoly, since only railroads and established companies could afford to purchase coal lands.[52] Yet, contrary to Mondell's fear, coal entries and sales increased. Immediately after the administration established the new policy, both entries and sales rose sharply, and then dropped off in 1911 and 1912.[53] Coal development on the public domain, it appears, depended on factors other than the new federal policy. Federal rates seemed quite low. In one large sale of valuable coal land in Wyoming, for example, the new system raised the price from $13,300 to $363,330 for 665 acres. Yet, since the area contained large deposits, the total purchase price came to less than ½ cent a ton.[54] Moreover, as with grazing fees, coal operators paid the federal government less per ton mined than they returned, in the form of royalties, to private and state land owners.[55] The Roosevelt administration did not respond favorably to Mondell's fear that corporations would replace individual enterprise in the coal fields. Roosevelt himself favored large-scale investment in the coal industry, which, he felt, would bring low-cost production.

The decision to lease water power sites only angered the West even more. Federal legislation, Westerners argued, had confirmed titles to water on the public domain which had been established under state law. The Forest Service, therefore, could not interfere with these titles by imposing a fee. The Forest Service tried to circumvent this argument by first adopting the legal theory of a "conservation charge," whereby the power company would pay for a water supply conserved by the national forests. Later the Service

[51] Pease, "Land Transgressor."

[52] Mondell to Fisher, Aug. 21, 1912 (GP #1672).

[53] Computed from the Annual Reports of the U. S. Geological Survey, 1900–1914.

[54] Fisher to Mondell, Aug. 17, 1912 (GP #1672).

[55] *Proceedings, National Conservation Congress, 1911*, 194–203.

shifted to the position that the fee constituted a simple land rent.[56] Representative Mondell again entered the fray, and in 1907 he unsuccessfully sought an amendment to the Agricultural Appropriation Bill to prevent such charges.[57] In 1909, again without success, he co-sponsored a measure to grant power permits in perpetuity, without charge. Despite the failure of Congress to act, Western politicians continued to work closely with water power corporations to obtain favorable legislation. In 1915 a special convention at Portland, Oregon, organized by the power companies, protested against legislation confirming the water power rentals devised by the Departments of Agriculture and Interior.[58] The following year the Supreme Court upheld the fee policy, and four years later Congress passed a comprehensive act regulating water power development. These actions silenced Western opposition and securely established federal control over water power on the public lands.

Organized Western Protest

As the Roosevelt natural resource program moved forward, more Western groups joined the outcry against the administration. Gradually they formed an organized opposition which presented a formidable challenge to the new policies. Prior to 1906–7, when the administration embarked upon its expanded public lands program, special legislation or administrative action had appeased one dissenting group after another. For example, in the Forest Administration Act of 1897, miners obtained free access to the forest reserves to prospect, enter, and patent mineral lands. Similarly, Pinchot had intervened on behalf of the stockmen to permit more extended grazing in the reserves. In fact, during the winter of 1904–5, the West seemed singularly friendly to the Roosevelt policies, as indicated by its support of the transfer of the reserves to the Department of Agriculture. But the decisions to charge for grazing, to lease coal lands and water power sites, and to control grazing on the public

[56] Wells, "Personal History" (GP #1671).

[57] *Forestry and Irrigation* (Apr. 1907), 13, 204.

[58] *Call, Resolutions, and Official Proceedings, Western States Water Power Conference* (Portland, 1915).

domain reversed this growing friendliness. These innovations produced the first organized regional reaction to the conservation program, and touched off a concerted movement to destroy it by ceding the public lands to the states.

The Denver Public Lands Convention in the summer of 1907 marked the first organized Western opposition. It came immediately upon the announcement of the new public lands policy in December 1906 in Roosevelt's message to Congress.[59] Confined primarily to Wyoming and Colorado, this movement was led by Representative Mondell of Wyoming and Senator Teller of Colorado. Of a total of 644 delegates at the meeting, 145 came from Wyoming and 386 from Colorado. The remaining 113 from other Western states, far more friendly to the administration, hesitated to join the attack.[60] The Denver convention, reflecting conflicting Western attitudes, could not take a strong stand against the administration. The public domain leasing proposition might have produced a unified opposition, and, in fact, the convention went on record against it. But the defeat of the leasing law in Congress even prior to the meeting lessened the impact of this protest. The obvious attempt of Wyoming and Colorado leaders to dominate the convention alienated many delegates from other states.[61] For the most part, therefore, the convention's resolutions dealt with minor administrative practices of the Forest Service.

The Wyoming and Colorado political leaders who had organized the Denver convention continued to criticize the administration through the newly formed Public Domain League. Bidding for popular support, the League issued press releases and bulletins which played up the difficulties encountered by settlers in entering farm land within the national forests. In commercial and political conventions they spoke for resolutions attacking Pinchot, the Forest Service, and the administration's entire resource policy. Invariably they proposed that Congress cede the public lands to the states.

[59] *Forestry and Irrigation* (June 1907), 13, 278.

[60] *Ibid.*, 279; Pease, "Land Transgressor"; Wells, "Personal History" (GP #1671).

[61] Pease, "Land Transgressor"; Peffer, *Public Domain*, 99–102.

When little support appeared for this change, they then demanded that all nonforest land be excluded from the reserves. These efforts agitated the West, but aroused little sympathy in Congress.[62]

The Taft administration also gave little encouragement to Western demands that federal lands be ceded to the states. In a conservation speech in the late summer of 1909, amid the brewing Pinchot-Ballinger fight, the President emphasized that he would not retreat from the principle of federal ownership of the public lands. Moreover, although he felt that Roosevelt illegally had inaugurated specific management policies, Taft did not disapprove of those same innovations if enacted into law by Congress.[63] Many Western groups hoped that Ballinger's appointment as Secretary of the Interior would bring about a wholesale reversal of resource policy. They were to be disappointed.[64] Ballinger did prefer that Congress grant the public lands to the states, but he obtained Taft's backing only in the single instance of the water power sites, and this, only temporarily.

The attack on federal resource policy became involved in partisan politics when Western Democrats took up the cause. Leading officials of the National Public Domain League, for example, were Democratic politicians. State leaders, such as governor Elias M. Ammons of Colorado, increased their personal political strength by exploiting hostility toward the Republican land policies.[65] Colorado and Wyoming politicians spearheaded the meeting of yearly regional Governors' Conferences to channel Western sentiment more effectively.[66] Pending favorable congressional action on ceding

[62] *Proceedings, Trans-Mississippi Commercial Congress, 1909*, 77, 83, 169–203, 249–281; W. G. M. Stone to Pinchot, Aug. 26, 1909 (GP #97); Clarence P. Dodge to Pinchot, June 29, 1909 (GP #120); NA, RG #48, file 2–132, National Public Domain League.

[63] *Proceedings, National Irrigation Congress, 1909*, 519–527.

[64] *Irrigation Age* (Aug. 1909), 24, 367.

[65] A. E. deRicqles to Pinchot, Feb. 11, 1913 (GP #1821).

[66] *Pueblo* (Colo.) *Chiefton*, Apr. 12, 1914 (GHM, Clippings, v. 18); Wells to Slattery, Aug. 29, 1913 (GP #1827); Frank D. Brown to Pinchot, Mar. 3, 1913 (GP #1821).

the public lands to the states, they urged each state to pass laws providing for effective management machinery which could later be placed into operation on federal lands.

Agitation against federal policies attracted considerable support in the West, but the campaign met stubborn resistance from other sections of the country. When in the summer of 1912 Westerners introduced in the Democratic National Convention a plank calling for granting the public lands to the states, they received little support.[67] Later that same month, Senator Overman of North Carolina introduced in the Senate a resolution calling for investigation of the Forest Service. This measure, calculated to create political ammunition for the Democrats, also failed to pass.[68] When they could not obtain congressional action on a state cession law, Western Democrats introduced measures to eliminate nonforested land from the national forests.[69] Pinchot and the National Conservation Association, working through the women's organizations, obtaining the support of railroad executives, and bombarding Congress with letters favorable to the Forest Service from those living in or near the national forests, aroused sufficient public protest to forestall this move.[70]

Western Democratic hopes rose when Wilson was elected President in the fall of 1912. Two months later the President-elect seemed to confirm these hopes when in a Chicago speech he remarked that "a policy of reservation is not a policy of conservation."[71] Triumphant Democrats from Colorado, led by their new governor, Elias M. Ammons, tried to persuade Wilson to appoint former governor

[67] Memorandum, Harry A. Slattery to Pinchot, July 5, 1912 (GP #1817).

[68] Slattery to Pinchot, July 23, 1912 (GP #1817).

[69] *Denver Republican*, Jan. 14, 1913 (GP #1826); *American Conservation* (Mar. 1911), I, 39, 67–68; (July 1911), I, 224–225.

[70] W. G. M. Stone to Slattery, May 27, 1912 (GP #1818); Mrs. James W. Pinchot to the Daughters of the American Revolution, nd (GP #1817); Overton W. Price to Pinchot, Oct. 30, 1912 (GP #1817); Henry S. Graves to Pinchot, Dec. 30, 1912 (GP #1819); Pinchot to James W. Bryan, June 9, 1913 (GP #1821); Howard Elliott to Pinchot, Jan. 25, 1913 (GP #1822).

[71] Robbins, *Our Landed Heritage*, 380.

Alva Adams of Colorado as Secretary of the Interior.[72] When the President intimated that he preferred to continue Walter Fisher in that post, Western Democrats, meeting in Washington, demanded that Wilson select a man from a public lands state. The appointment of Franklin K. Lane, a Californian, seemed to mollify the Westerners as well as to please Pinchot.[73] Late in January the former Chief Forester wrote: "All the information I get about Wilson's attitude toward our conservation questions is most satisfactory."[74] There seemed to be little danger that the new president would reverse the principle of federal ownership of the public lands. In May 1913 he declined to heed the advice of a special delegation of Colorado Democrats that he advocate cession of the national forests to the states,[75] and throughout his administration Wilson continued to disappoint his Western party leaders by remaining an advocate of a federal resource policy.

[72] Clarence P. Dodge to Pinchot, Nov. 16, 1912 (GP #1817); W. G. M. Stone to Pinchot, Jan. 11, 1912 (GP #1820); A. E. deRicqles to Pinchot, Feb. 11, 1913 (GP #1821); C. B. Rhodes to Pinchot, Jan. 18, 1913 (GP #1825).

[73] Memorandum, Jan. 30, 1913 in A. L. Pierce to unnamed person, Feb. 4, 1913 (GP #1827); D. W. Aupperle to Pinchot, Feb. 8, 1913 (GP #1823); Slattery to Pinchot, Jan. 17, 1913 (GP #1828).

[74] Pinchot to Henry Wallace, Jan. 31, 1913 (GP #1827).

[75] Slattery to Price, May 27, 1913 (GP #1828).

Chapter XIII

The Conservation Movement and the Progressive Tradition

The progressive revolt of the early twentieth century, so most historians have argued, was an attempt to control private, corporate wealth for public ends. The conservation movement typified this spirit. According to one such writer,

> . . . the progressive movement in the Republican party during Mr. Roosevelt's administration manifested itself primarily in a struggle against corporations. The struggle had two phases; first and most important, was the attempt to find some adequate means of controlling and regulating corporate activities; and second, and almost as important, was the resistance to the efforts of corporations to exploit the natural resources of the nation in their own behalf.[1]

Professor Roy Robbins, author of the most important one-volume history of the public lands, supports this view. He describes conservation as a popular reaction to the post-Civil War influence of private corporations in federal public land policy.[2] Gifford Pinchot's autobiography, moreover, contains strong antimonopoly overtones. He wrote: "Its [monopoly's] abolition or regulation is an inseparable part of the conservation policy."[3] This view has taken such deep root in the mind of the general public that conservation crusaders need only to expose the hand of corporate business to brand

[1] Stahl, "Ballinger-Pinchot Controversy," 69.

[2] Robbins, *Our Landed Heritage*, 301–423.

[3] Pinchot, *Breaking New Ground*, 507.

a resource measure as "anti-conservation" and detrimental to the public interest.[4]

Ownership or Use?

This point of view correctly describes the ideology of the conservation movement, but fails to analyze its broader meaning. It stresses conservation as a theory of resource ownership when, in fact, the movement was most concerned with resource use. To most historians the amount of exploitation in the nineteenth century varied directly with ownership. Large corporations, they argue, wasted resources lavishly, while small farmers did not. Professor Robbins writes: "The agency most responsible for this exploitation was not the individual farmer who typified the earlier period of American history, but the corporation which with abundant capital at its disposal was able to appropriate large areas of valuable land and often to exact an exorbitant tribute from the people who were attempting to build up the civilization of the country."[5] While utilizing antimonopoly overtones to make the reasoning seem plausible, this argument actually proceeds from premises as to land ownership to conclusions concerning land use.[6] To support his contention Robbins cites evidence that corporations obtained large areas of land from the federal government and widely used fraudulent land entries to enlarge their holdings. These facts help to establish the pattern of land ownership, but they do not pertain to problems of land use. As Thomas C. Chamberlin, professor of geology at the University of Chicago, wrote in 1910:

> In their fundamental nature, the problems of conservation and the problems of possession are distinct questions, each to be solved in its own way and on its own basis. . . . The conservation of natural resources

[4] See, for example, Arthur Carhart, "The Menaced Dinosaur Monument," *National Parks Magazine* (Jan.–Mar. 1952), 26, 19–30.

[5] Robbins, *Our Landed Heritage*, 301.

[6] *Ibid.*, 236–298. In a chapter on "exploitation under the settlement laws," pp. 236–254, Robbins seems to define "exploitation" more in terms of concentration of land ownership through fraudulent use of existing laws than in terms of wasteful land use.

centers in the scientific and technical; the right of ownership and the most desirable form of ownership center in the political and sociological. . . . [The] ownership or distribution of values . . . has no logical relation to conservation and may even be incompatible with its highest realization.[7]

And Philip P. Wells, former law officer of the United States Forest Service, complained in 1919 that too many writers described the conservation movement primarily as a protest against "land grabbing."[8]

Resource exploitation, in fact, reflected the attitude not merely of corporations, but of Americans in all walks of life. Small farmers, as well as corporate leaders, helped to establish a wasteful pattern of land use. Everyone in the nineteenth century hoped to make a killing from rising land values and from quickly extracting the cheap, virgin resources of the nation. Corporations often did exploit resources, such as the timber of the Great Lakes forest region. Such examples, however, do not support the general view that corporations by their very nature promoted resource waste, and the larger the corporation and the greater its self-interest, the more destruction it caused. On the contrary, when the conservation movement arose in the early twentieth century, it became clear that larger corporations could more readily afford to undertake conservation practices, that they alone could provide the efficiency, stability of operations, and long-range planning inherent in the conservation idea. Larger owners could best afford to undertake sustained-yield forest and range management, and understood more clearly than did small farmers the requirements for large-scale irrigation and water power development.

That large owners frequently supported conservation policies and small owners just as frequently opposed them also forces the historian to rethink the movement's significance. Many have argued that the Forest Reserve Act of 1891 and subsequent conservation measures reflected the antimonopoly movement of the post-Civil War Era. In explaining the "rise of conservation," Robbins writes:

[7] *American Forests* (Oct. 1910), 16, 10.
[8] Philip P. Wells to Pinchot, Jan. 3, 1919 (GP #1676).

Conviction that the federal government had been all too generous in its disposition of favors during the period between 1850 and 1870 produced a reaction in the form of an antimonopoly movement which demanded legislation to restore the public domain, and to provide equal opportunity for the many and special privilege to none. . . . Not until after the railroad magnate, the cattle king, the mining baron and the lumber monarch had established a prestige as great as that enjoyed by any capitalist of the eastern industrial order, did the federal government finally pass the first of a series of laws which was ultimately to be distinguished as the conservation movement.[9]

Yet, antimonopolists did not conspicuously push conservation measures, and in fact frequently opposed them. The campaign to establish forest reserves had its origin not in antimonopolism, but in the drive by wilderness groups to perpetuate untouched large areas of natural beauty, by Eastern arboriculturists and botanists to save trees for the future and by Western water users, both large corporations and small owners, to preserve their water supply by controlling silting. Large cattlemen backed the range leasing measure, while settlers opposed it. Groups representing small farmers supported Pinchot during his fight for federal water power regulation, but this support came from those whose lands were threatened with flooding by proposed private water power reservoirs, and who in later years repeatedly opposed federal reservoir construction for the same reason.

The movements for wider land distribution and more efficient land use had entirely separate origins. The first continued after the Forest Reserve Act of 1891, but in a direction diametrically opposed to the spirit of that law. Homesteaders bitterly resented permanent reservation of public land from private entry. To them, forest, range, or mineral reserves differed little from withdrawals for railroad land grants. They fought with equal vigor to abolish both. At the same time, conservation leaders felt closer to the spirit of development, typified by the railroad land grants, than to the reaction against the roads. Both railroads and conservationists promoted large-scale economic development. While the conservation move-

[9] Robbins, *Our Landed Heritage*, 301.

ment emphasized greater efficiency in this process, its goal of planned economic growth and its consolidating tendencies closely approximated the spirit of railroad construction. The transcontinental lines, in fact, cooperated closely with conservationists in developing Western resources, and gave special aid to federal irrigation, forest, and range programs.

In placing the conservation movement in the context of progressive ideology, historians have concentrated on incidents which easily fit that viewpoint and have avoided problems which it could not easily explain. They have analyzed extensively such events as the Pinchot-Ballinger controversy, Teapot Dome, the fight for water power regulation, and the growth of federal forest and oil policy. These incidents they have interpreted as struggles between private and public ownership. They have associated private ownership with the "corporation" and resource exploitation, and public ownership with the "people" and wise resource use. Historians, however, have undertaken no corresponding investigations of federal irrigation or river development. Federal range control, a crucial issue which displayed an alignment of forces exactly opposite of those indicated by the progressive ideology, received little attention until 1951.[10] These problems played as vital a role in the Theodore Roosevelt conservation movement as did forest, mineral, and power policies. Traditional views of conservation which do not account for these developments must give way to newer interpretations.

The broader significance of the conservation movement stemmed from the role it played in the transformation of a decentralized, nontechnical, loosely organized society, where waste and inefficiency ran rampant, into a highly organized, technical, and centrally planned and directed social organization which could meet a complex world with efficiency and purpose. This spirit of efficiency appeared in many realms of American life, in the professional engineering societies, among forward-looking industrial management leaders, and in municipal government reform, as well as in the resource management concepts of Theodore Roosevelt. The possibilities of applying scientific and technical principles to resource

[10] Peffer, *Public Domain*.

development fired federal officials with enthusiasm for the future and imbued all in the conservation movement with a kindred spirit. These goals required public management, of the nation's streams because private enterprise could not afford to undertake it, of the Western lands to adjust one resource use to another. They also required new administrative methods, utilizing to the fullest extent the latest scientific knowledge and expert, disinterested personnel. This was the gospel of efficiency—efficiency which could be realized only through planning, foresight, and conscious purpose.

The lack of direction in American development appalled Roosevelt and his advisers. They rebelled against a belief in the automatic beneficence of unrestricted economic competition, which, they believed, created only waste, exploitation, and unproductive economic rivalry. To replace competition with economic planning, these new efficiency experts argued, would not only arrest the damage of the past, but could also create new heights of prosperity and material abundance for the future. The conservation movement did not involve a reaction against large-scale corporate business, but, in fact, shared its views in a mutual revulsion against unrestrained competition and undirected economic development. Both groups placed a premium on large-scale capital organization, technology, and industry-wide cooperation and planning to abolish the uncertainties and waste of competitive resource use.

Theodore Roosevelt and the Conservation Movement

Historians of the Progressive Era have found it increasingly difficult to categorize Theodore Roosevelt. Was he a "liberal" or an "enlightened conservative?" Did he rob the Democrats of their reform proposals and fulfill the aims of late nineteenth-century social revolt, or did he merely mouth their causes and, in practice, betray them? These questions pose difficulties chiefly because they raise the wrong issues. They assume that the significance of Roosevelt's career lies primarily in its role in the social struggle of the late nineteenth and early twentieth centuries between the business community on the one hand and labor and farm groups on the other. On the contrary, Roosevelt was conspicuously aloof from that social struggle. He refused to become identified with it on either side. He was, in fact, predisposed to reject social conflict, in theory and practice, as the greatest danger in American society. His administration and his social and political views are significant primarily for their attempt to supplant this conflict with a "scientific" approach to social and economic questions.

Roosevelt was profoundly impressed by late nineteenth-century social unrest, and in particular by its more violent manifestations such as the Haymarket riot, the Pullman strike, Coxey's army, and the election of 1896. He viewed Populism as a class struggle which would destroy the nation through internal conflict. He rejected Western Insurgency—a continuation of Populist radicalism, he believed—because it expressed the aims of only one economic group in society which, if dominant, would exercise power as selfishly as did the Eastern business community. The economic struggle slowly evolving from rapid industrial growth aroused Roosevelt's deepest fears. His practical solution to that problem consisted neither of granting dominance to any one group, nor of creating a balance of power among all. In fact, as a result of his fear of conflict he almost denied its reality and tried to evolve concepts and techniques which would, in effect, legislate that conflict out of existence.

Social and economic problems, Roosevelt believed, should be solved, not through power politics, but by experts who would undertake scientific investigations and devise workable solutions. He had an almost unlimited faith in applied science. During his presidency, he repeatedly sought the advice of expert commissions, especially in the field of resource policy, and he looked upon the conservation movement as an attempt to apply this knowledge. But he felt that government could tackle nonresource questions such as labor problems with the same approach. In the fall of 1908 he wrote to a union official:

Already our Bureau of Labor, for the past twenty years of necessity largely a statistical bureau, is practically a Department of Sociology, aiming not only to secure exact information about industrial conditions but to discover remedies for industrial evils. . . . It is our confident claim . . . that applied science, if carried out according to our program, will succeed in achieving for humanity, above all for the city industrial worker, results even surpassing in value those today in effect on the farm.[11]

Having little appreciation of labor's permanence as a power group in society, Roosevelt believed simply that one could approach the labor problem by improving working conditions, training more efficient employees, and stimulating the industrial machine to prevent unemployment.

President Roosevelt's abiding fear of class struggle led him to conceive of the good society as a classless society, composed, not of organized social groups, but of individuals bound together by personal relationships. Believing that "the line of division in the deeper matters of our citizenship" should "be drawn on the line of conduct," he viewed the fundamentals of social organization as personal moral qualities of honesty, integrity, frugality, loyalty, and "plain dealing between man and man." Thus, he was "predisposed to interpret economic and political problems in terms of moral principles." These moral qualities resided, not in the urban centers, which bred only social disorder, but among the "farmer stock" which possessed "the qualities on which this Nation has had to draw in order to meet every great crisis of the past."[12] Agricultural life was the best means of obtaining that "bodily vigor" which produces "vigor of soul," and independent, property-owning farm families were the major source of social stability and the bulwark against internal conflict. For Roosevelt, with his interest in the out-of-doors, his emphasis on moral vigor arising from struggle with the elements, and his basic fear of social unrest, the good society was agrarian. He would have opposed bitterly any effort to turn back the industrial clock, yet his ultimate scheme of values was firmly rooted in an agrarian social order.

Roosevelt's emphasis on applied science and his conception of the good society as the classless agrarian society were contradictory trends of thought. The one, a faith which looked to the future,

[11] Roosevelt to P. H. Grace, Oct. 19, 1908 (TR).

[12] Roosevelt to Taft, Dec. 21, 1908 (TR).

accepted wholeheartedly the basic elements of the new technology. The other, essentially backward-looking, longed for the simple agrarian Arcadia which, if it ever existed, could never be revived. He faced two directions at once, accepting the technical requirements of an increasingly organized industrial society, but fearing its social consequences. In this sense, and in this sense alone, Roosevelt sought Jeffersonian ends through Hamiltonian means. He had great respect for both men, each of whom manifested one side of his own contradictory nature. But he admired even more Abraham Lincoln, the spokesman of the "plain people," whose life combined agrarian simplicity and national vigor. By the same token, Roosevelt considered his irrigation program as one of his administration's most important contributions. It expressed in concrete terms his own paradoxical nature: the preservation of American virtues of the past through methods abundantly appropriate to the present.

The contradictory elements of Roosevelt's outlook fused also in an almost mystical approach to the political order best described as "social atomism." Strongly affirming the beneficial role of both expert leadership and the vast mass of humanity, he could not fit into his scheme of things intervening group organization on the middle levels of power. Americans should live, he thought, as individuals rather than as members of "partial" groups, their loyalties should be given not to a class or section but to their national leader. As his administration encountered continued difficulty with Congress, Roosevelt relied more and more on executive commissions, and on action based upon the theory that the executive was the "steward" of the public interest. Feeling that he, rather than Congress, voiced most accurately the popular will, he advocated direct as opposed to representative government. Unable to adjust to a Congress which rejected his gospel of efficiency, Roosevelt took his case to the "people." In doing so he not only bypassed the lawmakers but also defied the group demands of organized American society. Growing ever more resentful of the hindrances of a Congress which expressed these demands, Roosevelt drew closer to a conception of the political organization of society wherein representative government would be minimized, and a strong leader, ruling through vigorous purpose, efficiency, and technology, would derive his support from a direct, personal relationship with the people.

As president, Roosevelt concentrated on problems which would not raise issues of internal social conflict—foreign policy and conservation. Increasingly stressing the conservation program during his second term of office, Roosevelt looked upon it as the most important contribution of his administration in domestic affairs. Conservation gave wide scope to government by experts, to investigation by commissions, to efficiency in planning and execution. It called forth patriotic sentiments which could override internal differences. More efficient production of material goods would help solve the labor problem in the way he thought it could be solved—by providing full employment and lower living costs. Even more important, through the federal irrigation program and the Country Life Movement, both of which Roosevelt encouraged, the President thought that he was buttressing the "Republic" in its most vital spot. Warning President-elect Taft that rural migration to cities would create a decline in the nation's population, in December 1908 Roosevelt urged his successor to formulate a program for country life improvement. "Among the various legacies of trouble which I leave you," he entreated, "there is none to which I more earnestly hope for your thought and care than this."[13]

Herbert Croly's *The Promise of American Life*, written in 1908, articulated these tendencies in Roosevelt's political and social thought. Croly deeply feared that group consciousness in America would lead at best to an aimless, drifting society, and at worst to disastrous internal conflict. Vigorous, national purpose, he argued, should replace the current American faith in automatic evolution toward a better society. Less a blueprint than a simple plea for action, *The Promise of American Life* immediately appealed to Roosevelt as the scholarly expression of the assumptions upon which he had acted as president. And in domestic affairs there was no better illustration of those assumptions than the administration's conservation policies. Croly's work, the president declared, was "the

[13] Roosevelt to Taft, Dec. 21, 1908 (TR).

most profound and illuminating study of our national conditions which has appeared for many years."[14]

In holding these attitudes, Roosevelt personally embodied the popular impulses which swung behind the conservation movement during the years of the great crusade. That crusade found its greatest support among the American urban middle class which shrank in fear from the profound social changes being wrought by the technological age. These people looked backward to individualist agrarian ideals, yet they approved social planning as a means to control their main enemy—group struggle for power. A vigorous and purposeful government became the vehicle by which ideals derived from an individualistic society became adjusted to a new collective age. And the conservation movement provided the most far-reaching opportunity to effect that adjustment. Herein lay much of the social and cultural meaning of the movement for progressive resource planning.

Conservation and the Grass Roots

The deepest significance of the conservation movement, however, lay in its political implications: how should resource decisions be made and by whom? Each resource problem involved conflicts. Should they be resolved through partisan politics, through compromise among competing groups, or through judicial decision? To conservationists such methods would defeat the inner spirit of the gospel of efficiency. Instead, experts, using technical and scientific methods, should decide all matters of development and utilization of resources, all problems of allocation of funds. Federal land management agencies should resolve land-use differences among livestock, wildlife, irrigation, recreation, and settler groups. National commissions should adjust power, irrigation, navigation, and flood control interests to promote the highest multiple-purpose development of river basins. The crux of the gospel of efficiency lay in a rational and scientific method of making basic technological decisions through a single, central authority.

[14] Roosevelt, "Nationalism and Popular Rule," *The Works of Theodore Roosevelt* (National Edition) (20 v., New York, 1926), v. 17, 53.

Resource users throughout the country differed sharply from this point of view. They did not share the conservationists' desire for integrated planning and central direction. Instead, each group considered its own particular interest as far more important than any other. Resource users formed their opinions about conservation questions within the limited experience of specific problems faced in their local communities. They understood little and cared less for the needs of the nation as a whole. This approach to resource affairs stood in direct contrast to the over-all point of view of the conservationists. While the first gave rise to centrifugal tendencies in resource management, the second produced centripetal influences. While resource-use groups held a multitude of diverse aims which stemmed from many limited and local experiences of particular problems, conservationists held comprehensive and unified objectives. An expert adjustment of resource conflicts plainly would fulfill the broad objectives of conservation leaders, but it hardly sufficed to achieve the aims of particular localities more concerned with the problems which they knew firsthand.

Roosevelt conservation leaders had difficulty in adjusting these conflicting outlooks. Their entire program emphasized a flow of authority from the top down and minimized the political importance of institutions which reflected the organized sentiment of local communities. Pinchot and Roosevelt did take into account grass-roots interests, but only to facilitate administration and to prevent their decisions from arousing too much resentment. They postponed forest reserve executive orders until after elections, and cultivated the favor of stockmen to gain support for the transfer of the reserves to the Department of Agriculture. The Forest Service decentralized its administration so that federal officials could become better acquainted with local interests. Conservation leaders, however, rarely, if ever, permitted grass-roots groups to decide policy questions. These matters, they argued, could be left to local groups, or to political pressure in Congress only at the risk of "selling out" the national welfare to "special interests."

Grass-roots groups throughout the country had few positive objectives in common, but they shared a violent revulsion against the scientific, calculated methods of resource use adjustment favored by the conservationists. Both large and small property owners knew that the conservationists' plans involved methods of decision far beyond their control, and each group feared that a broader program would obscure its own specific needs or minimize its own project. Basin-wide river planning might require a dam in another locality. Multiple-purpose dams might provide less water desperately needed for navigation and more for electric power for some remote industry. Rigid grazing control might benefit the irrigator in the lower basin, but curtail the activities of stockmen on the headwaters. Each group desired financial and technical aid from the federal government, and each supported executive action when favorable to it, but none could feel a deep sense of participation in the process by which technical experts made resource decisions. Experience with the Forest Service and Bureau of Reclamation alienated many groups which found it difficult to influence administrative policy. They opposed plans to establish executive adjustment of conflicting uses and favored methods of decision over which they felt they had some measure of control.

Grass-roots groups utilized a variety of political methods, both judicial and legislative, to protect their interests. Through pressure on federal agencies or influence in selecting personnel, they could even modify the administrative process itself. Western water users, for example, resorted to the courts to counteract a rational state water law and efficient federal water development, both of which they viewed as a menace to their existing rights. The water laws proposed by Elwood Mead provided for administrative determination of available supplies and existing priorities, adjustment of present conflicts, and supervision of future filings. Fearing that their claims would suffer under this arrangement, water users preferred the older method of judicial determination of rights, a procedure in which they felt they had some degree of influence. Water users involved in interstate conflicts preferred to present their cases to the federal courts rather than to permit an administrative determination of rights which might follow the requirements of development rather than the merits of each individual claim.

Resource groups frequently obtained crucial influence with federal agencies which dealt with their problems. Not until 1913 did water users move into the inner circles of the Bureau of Reclamation, but after 1901 Western stockmen held a key position in the Forest Service. Using Albert Potter as his major contact with the grazers, Pinchot developed a close working arrangement with the large cattlemen to obtain their support for the transfer. Throughout the years before World War I the same group continued to support the Forest Service, through a tacit agreement that if federal officials would push a leasing program for the public domain, the cattlemen would not object to administrative grazing regulations. This happy arrangement then came to an end, as decisions to increase grazing fees and to reduce the number of livestock permitted in the forests both revealed the declining influence of the stockmen and precipitated their open hostility toward the Forest Service.

Nationwide pressure groups became the most effective technique adopted by resource users to influence resource decisions. Organizations such as the National Rivers and Harbors Congress, the American National Livestock Association, or the National Water Users Association grew up to represent the active segments of particular interests in their bid for influence in the conservation program. Although they quickly obtained political support from local congressmen, they remained thoroughly nonpartisan. Cooperating closely with congressional committees which dealt with special resource problems, each group bargained politically with others to obtain sufficient votes to pass its program. Through these logrolling techniques Congress developed many projects at once to satisfy a great number of localities rather than to construct the most important ones first. Through the same method it preferred to scatter appropriations over many projects to be spent year by year rather than to concentrate them on those developments most needed. These methods of national political organization and bargaining constituted the characteristic pattern of making resource decisions. Through them local groups achieved a sense of participation in and control over resource development which they did not receive from the more centrally directed methods of the conservationists.

Resource users played a fundamental role in shaping the character of development in a manner contrary to the aims of conservationists. They created a single, rather than a multiple-purpose attack on resource affairs. Economic organizations concerned with single interests—such as navigation, flood control, or irrigation—joined with administrative agencies in charge of individual programs and congressional committees which dealt with specialized subjects to defeat an integrated approach. Through policies devoted to the development of a single resource, Congress found protection against independent executive action, administrative agencies discovered a means to prevent coordination of their work with other bureaus, and local interests created programs of direct benefit to themselves and under their control. Private organizations and their congressional allies established this pattern. Although administrative agencies, such as the Corps of Engineers, took much initiative in preserving their administrative independence, their concern for single-purpose development reflected rather than molded the attitude of Congress. Single-purpose policies, impractical from the point of view of the conservation ideal of maximum development through scientific adjustment of competing uses, became the predominant pattern because they provided opportunities for grass-roots participation in decision-making. They enabled resource users to feel that they had some degree of control over policies that affected them.

The first American conservation movement experimented with the application of the new technology to resource management. Requiring centralized and coordinated decisions, however, this procedure conflicted with American political institutions which drew their vitality from filling local needs. This conflict between the centralizing tendencies of effective economic organization and the decentralizing forces inherent in a multitude of geographical interests presented problems to challenge even the wisest statesman. The Theodore Roosevelt administration, essentially hostile to the wide distribution of decision-making, grappled with this problem but failed to solve it. Instead of recognizing the paradoxes which their own approach raised, conservationists choose merely to identify their opposition as "selfish interests." Yet the conservation movement raised a fundamental question in American life: How can large-scale economic development be effective and at the same time fulfill the desire for significant grass-roots participation? How can the technical requirements of an increasingly complex society be adjusted to the need for the expression of partial and limited aims? This was the basic political problem which a technological age, the spirit of which the conservation movement fully embodied, bequeathed to American society.

Bibliographical Note

The conservation movement has given rise to an enormous amount of literature during the past fifty years. Most of this published material, however, supports the traditional analysis of conservation as a struggle between the "interests" and the "people," and includes little evidence essential for a re-evaluation of conservation history. The best-known expression of the prevailing theory is contained in the last five chapters of Roy M. Robbins, *Our Landed Heritage* (Princeton, 1942), while a more recent restatement is J. Leonard Bates, "Fulfilling American Democracy: The Conservation Movement, 1907 to 1921," in *Mississippi Valley Historical Review* (June 1957), 44, 29–57.

The point of view of *Conservation and the Gospel of Efficiency* rests largely upon manuscripts produced by those who participated in the events between 1890 and 1920. These fall into four major categories: personal papers, records of federal departments, publications of organized interest groups, and official public documents. Each of these kinds of source material contributed in its own distinct manner to the larger picture of conservation history.

Many of the conservation leaders of the Theodore Roosevelt administration left personal manuscripts, some of them in enormous quantity, and most of which I was able to examine. The Pinchot and Newlands papers were the richest mines of information. The Gifford Pinchot manuscripts, located at the Library of Congress, are staggering in volume—some 3,300 file boxes of material. But an unusually complete index greatly simplifies the work of the researcher, and easily guides him to the some 500 boxes of material pertinent to natural resource activities prior to the First World War. These include, as well as Pinchot's own personal correspondence, the complete files of the National Conservation Association, a large collection of papers from the personal files of Philip P. Wells, a similar collection from the files of Eugene Bruce, a great number of extremely valuable histories of personal activities in conservation which Pinchot solicited early in the 1920's from men long active in the United States Forest Service, and numerous newspaper clippings from a variety of papers.

An equally valuable source of information, though not as large in quantity, are the Francis G. Newlands manuscripts located at the Sterling Memorial Library at Yale University. This collection includes some seventy boxes of letters, several hundred books, reports, and documents dealing with water problems, some one hundred scrapbooks of newspaper clippings from the entire country, a large collection of pamphlets and magazines on water development, and eight filing-cabinet drawers of material on conservation topics which the collection's custodian, Mr. Arthur B. Darling, has helpfully organized. Many of these manuscripts are available in printed form in Mr. Darling's edition of *The Public Papers of Francis G. Newlands* (2 v., New York, 1932). The Newlands papers proved of special value in shedding new light on the long-neglected drama of water conservation. The Newlands papers dovetailed at many points with the George H. Maxwell collection, located in the records of the Bureau of Reclamation at the National Archives. This large collection of letters, periodicals, and newspaper clippings contains little from the years of the fight for the Reclamation Act of 1902; however, it is extremely helpful for the period after 1905, and especially for the struggle for the Newlands waterway measure. The collection gives an illuminating picture of Maxwell as a promoter of a great variety of public causes.

Many other collections of personal manuscripts contributed less in quantity, but much that was vital. The Theodore Roosevelt collection at the Library of Congress, of which copies of outgoing letters are located also at Widener Memorial Library, Harvard University, yielded important information amid a voluminous output of presidential correspondence on a variety of topics. The William Howard Taft papers, at the Library of Congress, also contributed significant data both for his years as Secretary of War and as President. The manuscripts of WJ McGee, Frederick H. Newell, and James R. Garfield, also on deposit at the Library of Congress were of great value. And the papers of Myra Lloyd Dock, a Pennsylvania associate of Pinchot's, interested primarily in preservation, contributed additional insight into that point of view. I was extremely fortunate in being able to interview one of the leading figures of the multiple-purpose river development movement, Marshall O. Leighton, who in 1951 was still a consulting engineer in Washington.

The National Archives at Washington, D.C. contributed fully as much to this study as did the personal manuscripts. Here are goldmines of information, extensive and highly rewarding material for the researcher who has the fortitude to plow through the interminable number of boxes. Five of the Record Groups at the Archives were especially rich in material about the conservation movement: Record Group #48, Records of the Department of the Interior, Correspondence of the Office of the Secretary; Record Group #49, Records of the General Land Office; Record Group #77, Records of the War Department, Army Section, Correspondence of the Chief of the Army Corps of Engineers; Record Group #95, Records of the United States Forest Service; and Record Group #107, Records of the War Department, Army Section, Correspondence of the Secretary of War. These manuscripts provided invaluable insight into the attitudes and activities of federal resource agencies. They were especially helpful in establishing a view of policy-in-the-making and of the personal and group struggle over the formation and execution of policy. They provided the inside picture which the official documents often obscured.

Some material pertaining to the role of organized private interest groups can be found in the records of the federal agencies. But for most of the story one must go to the publications of the groups themselves, either the official proceedings of their conventions, or their periodicals. I have used the *Proceedings* of the conventions of a great number of such groups, including the American National Livestock Association, the American Society of Civil Engineers, the

International Dry Farming Congress, the Lakes-to-the-Gulf Deep Waterway Association, the National Conservation Congress, the National Irrigation Congress, and the Trans-Mississippi Commercial Congress. Regular publications of such organizations or of groups closely related to them also yielded considerable material; these include *American Conservation*, *American Forestry*, *Conservation*, *Engineering News, Engineering Record, Forest and Stream, The Forester*, *Forestry and Irrigation*, *Forward St. Louis*, *Irrigation Age*, *Maxwell's Talisman*, and the *Waterways Journal*.

Still a fourth group of materials important for this study were the official executive and legislative documents of the time. These vary in usefulness. Some, such as the annual reports of the various departments, are difficult to interpret unless one can view the struggle over policymaking through the personal and departmental manuscripts as well. Others, such as the reports of the Bureau of Corporations, contain valuable compilations of data and information which the historian could not possibly gather by himself. Especially fruitful as a source of information are the hearings and investigations conducted by Congress, although these are less helpful for the period before 1920 than after. For example, the *Investigations of the Department of the Interior and the Bureau of Forestry, Senate Document #719*, 61st Congress, 3rd Session, contains much information about conflicts over conservation policy unobtainable elsewhere, while Max W. Ball, *Petroleum Withdrawals and Restorations Affecting the Public Domain, United States Geological Survey Bulletin #623* (Washington, 1916) contains a valuable collection of documents about federal oil policy prior to World War I. The official documents used for this study are too numerous to detail here; a complete list can be found in the original copy of my doctoral dissertation at Widener Memorial Library, Harvard University.

The same can be said of the many books and articles on conservation, only a few of which can be listed here. Among the primary works, Gifford Pinchot's autobiography, *Breaking New Ground* (New York, 1947), though limited in perspective and viewing the conservation movement through his own eyes, is of considerable value. Theodore Roosevelt's *Autobiography* (New York, 1913), the conservation chapters of which were written, it appears, by Pinchot and Garfield, is also helpful. William B. Greeley, in *Forests and Men* (New York, 1951) gives a personal account of the history of forestry by one who was deeply involved in many phases of the pre-1920 movement. In *Constructive Democracy* (New York, 1905), William E. Smythe, a prominent Westerner, presents a highly illuminating point of view as to the need for positive government in resource development.

A number of secondary works on the conservation movement have been of special value to me. Of these, only two pertain to water development: Arthur Dewitt Frank, *The Development of the Federal Program of Flood Control on the Mississippi River* (New York, 1930), and Jerome Kerwin, *Federal Water-Power Legislation* (New York, 1926). The public lands have received far greater attention. The only major work to deal with the conservation movement on the public lands as a whole is Roy Robbins, *Our Landed Heritage* (Princeton, 1942), but there are others on more specific topics. Two books by John Ise, *The United States Forest Policy* (New Haven, 1920), and *The United States Oil Policy* (New Haven, 1926) were pioneering ventures. The best account of the public range is in H. Louise Peffer's *The Closing of the Public Domain* (Stanford, 1951), while the activities of preservationists can be followed best in William T. Hornaday, *Thirty Years War for Wildlife* (New York, 1931), and Robert Shankland, *Steve Mather of the National Parks* (New York, 1951). Two works, Rose M. Stahl, "The Ballinger-Pinchot Controversy," *Smith College Studies in History* (January 1926), XI (2), and Alpheus T. Mason, *Bureaucracy Convicts Itself: The Ballinger-Pinchot Controversy of 1910* (New York, 1941) deal with the Pinchot-Ballinger quarrel, but fail to grasp the basic points at issue or the larger significance of the affair.

Several doctoral dissertations have provided information, unavailable in other secondary works, about certain phases of conservation. The most valuable of these were Peggy Heim, "Financing the Federal Reclamation Program, 1902 to 1919: The Development of Repayment Policy" (Columbia University, 1953); Edward Lawrence

Pross, "A History of Rivers and Harbors Appropriation Bills, 1866–1933" (Ohio State University, 1938); Elmo R. Richardson, "The Politics of the Conservation Issue in the Far West," 1896–1913," (University of California, Los Angeles, 1958); and Lawrence Rakestraw, "A History of Forest Conservation in the Pacific Northwest, 1891–1913" (University of Washington, Seattle, 1955).

Numerous magazine articles and pamphlets complete the important sources for this study. Popular journal articles, for example, described many features of the enthusiasm for waterways. Among these are John L. Mathews, "The Future of Our Navigable Waters," *Atlantic Monthly* (December 1907), 100, 721–728, and Charles M. Harvey, "The Lakes-to-the-Gulf Deep Waterways Association," *World To-day* (January 1907), 12, 39–41. Others dealt with Western land problems, for example, Lute Pease, "The Way of the Land Transgressor," *Pacific Monthly* (August 1907), 18, 145–164, and William E. Smythe, "The Battle in the States," *Out West* (August 1902), 233–237. Two important articles of a more general nature are WJ McGee, "Water as a Resource," in "Conservation of Natural Resources," *Annals of the American Academy of Political and Social Science* (May 1909), 33, 37–50, and Charles R. Van Hise, "The Future of Man in America," *The World's Work* (June 1909), 18, 11718–11724. A variety of periodicals such as *American Review of Reviews*, *Public Opinion*, *North American Review*, *Survey*, *Century Magazine*, *Cassier's Engineering Magazine*, *Collier's Weekly*, and *California Illustrated Magazine* often contain articles of a similar nature.

Index